Derek Hough

EVOLUTION:

a case of stating the obvious

BERKELEY PUBLISHING
1997

Published by
Berkeley Publishing
PO BOX 7
Berkeley
Gloucestershire
GL13 9YH

First published in 1997 in paperback edition

ISBN 0-9529358-0-5

Printed in Great Britain by
Bailey Print Limited
Reliance House
Long Street
Dursley
Glos GL11 4LS

Cover taken from illustration by
Jay Edward Clements
"EVOLUXTAPOSITION"

ACKNOWLEDGEMENTS

The author kindly acknowledges the permission of Michael Denton to include the various extracts from his book *Evolution: a Theory in Crisis*

The quotations taken from *The Great Evolution Mystery* by Gordon Rattray Taylor are reprinted by permission of The Peters Fraser & Dunlop Group Ltd.

Suggestions for improvements to the text made by Christopher White were gratefully taken up by the author.

Contents

Preface

It is often difficult to analyse the rationale which lies behind holding a particular view or having a particular opinion. Some decisions are based on unambiguous evidence which leads to an obvious choice whereas other decisions can be based on numerous finely balanced arguments and our final choice between the alternatives can become what is known as a "gut feeling." This is often the result of an almost sub-conscious intellectual process helping to make our minds up. While the decision itself is passed to the conscious mind the exact route taken to that decision remains obscure and difficult to recognise. After years of dedicated belief in Darwin's Theory it was this sub-conscious distilling of information which suddenly led me to reject Darwinism as being a comprehensive explanation of evolution. This book is an attempt to explain the reasoning which led to this change of mind.

Every year many books are written on evolutionary subjects, some in favour, and some anti-Darwinian. I have quoted freely from the two books which, more than any others, motivated me to rethink my commitment to Darwin's Theory. They are Michael Denton's *Evolution: A Theory in Crisis* and Gordon Rattray Taylor's *The Great Evolution Mystery*. These books, which have been written by professional scientists, point out many of the weaknesses of the accepted theory of evolution and although the evidence against Darwinism is brilliantly argued neither book strongly suggests an alternative theory. The third book which also reinforced my doubts was, surprisingly, Darwin's original work on the subject, *The Origin of Species*. We shall see later that Darwin himself was less committed to Darwinism than are most twentieth century biologists.

The central theme of this book is to argue for an alternative mechanism to Darwin's natural selection as the main driving force of evolution. This work is not meant to be part of any anti-Darwinian crusade but its central argument is to emphasise the main weaknesses of Darwin's Theory. The book is aimed specifically at the layman and it is hoped that the reader will be convinced, as I am, that the argument put forward merely states the obvious.

Chapter One

Old Theories and New Definitions

The journey between Piccadilly Circus and Charing Cross on the London Underground takes approximately two minutes. The only bearing which this useless piece of information has on this story is that it was between these two stations that I rejected Darwin's Theory of Evolution and proposed an alternative theory which forms the subject matter of this book. Darwin's Theory was rejected as the train was pulling out of Piccadilly Circus station and in the final ninety seconds of the journey I pondered on other possible mechanisms which could explain the great variety of life which we observe in nature.

The reader may not be too impressed with a theory that took less than two minutes to formulate but anyone who has experienced the rush-hour conditions on the London Underground will know only too well how the mind is concentrated during those appalling journeys. It is an inhumane way to travel, especially for the English who place a very high value on privacy and territory. The passengers are packed like dried dates. Nobody speaks to or even looks at anybody else and the sanctity of one's twelve inch by twelve inch space is protected by losing oneself deep in private meditation.

I had been thinking about Darwin's Theory of Evolution continuously for the previous seventeen years on an almost daily basis. Darwin's Theory is not something that can be learnt, understood and accepted as a *fait accompli*. There is in fact no definite proof of the theory (it is difficult to conduct experiments over millions of years!) and one must continually check the theory against what we observe in nature to convince ourselves that the theory is adequate to explain all the complexities of life.

I consider Darwin's Theory to be the most intellectually seductive idea in scientific thought. When Darwin's book, *The*

3

Origin of Species, was published in 1859 contemporary scientists such as Thomas Huxley were immediately convinced by the theory and saw it as being so obviously correct. I was also initially convinced by this wonderfully simple explanation of the diversity of nature. In fact for seventeen years I was an almost fanatical supporter of Darwin and always took great delight in explaining evolution to interested (and uninterested) friends.

The great step forward which I took between Piccadilly Circus and Charing Cross was not to think up another mechanism of evolution but to reject Darwin's Theory as the main driving force of evolution. Once the step is taken of rejecting this mesmeric theory then it is a relatively simple matter to think up alternative theories. The hold which Darwin's Theory has on the academic world has largely been responsible for restricting discussion on other possible mechanisms. A large portion of academic energy is given over to defending Darwin's Theory which has been under constant attack since the day that *The Origin of Species* was published.

This book is aimed at the interested layperson and it is with him or her in mind that the next section is given over to an explanation of what Darwin and his followers believe to be the mechanism of evolution. It is important to have a thorough understanding of Darwin's Theory before we can reject it and move onto the new theory. In fact Darwin's Theory continues to play an important part in the new theory but not as the main mechanism of evolution.

Well, what exactly does Darwin's Theory say? It should first be pointed out that Darwin's original theory has been greatly improved upon with the help of twentieth century biological research. The study of genetics coupled with Darwin's original theory have come to be known as neo-Darwinism. For the sake of simplicity I shall continue to refer to "Darwin's Theory" even though most of the knowledge supporting the theory is of the twentieth century.

In Darwin's day the idea that evolution had occurred

was not a new one. The vast collections of fossils held by learned men and institutions led some scientists to the view that one species of organism could gradually change into another species. With the advent of the study of geology the fossils could be arranged in time sequence and this demonstrated that over long periods of time new species were being created and old ones were becoming extinct.

The great challenge facing Darwin was to explain how evolution could happen and to discover the driving force responsible for transforming one species into another.

Darwin knew that in a large population of any species there was a significant degree of variation between individuals and he also realised that far more offspring were produced each generation than could survive to maturity. He concluded that the offspring which tended to die before maturity were in some way less suited to the conditions in which they lived than those who did survive. It was well known that offspring tended to have similar characteristics to their parents and therefore the survivors in a large population would pass on their particular characteristics to the next generation. The characteristics of the survivors and therefore their offspring would in some way be more suitable for the conditions in which they lived than the characteristics of the individuals who did not survive. This same process of weeding out the less "fit" would go on generation after generation eventually changing the average characteristics of the individuals to make the population as a whole better suited to its environment.

This mechanism of eliminating the less fit relies on the availability of variation between individuals in the population. This variety comes about, according to twentieth century genetics, by something known as chance or random mutations of the genes. The genes are the chemicals of heredity which are passed from generation to generation by either simple cell duplication or by sexual reproduction and which give each individual organism its particular characteristics. A random mutation of the genes is a change in the chemical composition of a gene resulting from an error in the duplication process and this mutation will give the offspring a characteristic which is not inherited from the parents.

The environment in which the organism lives can be defined as any stimulus acting on the individual members of the population and it is pressure from this environment which selects out those individuals who should survive and pass on their more suitable genes to the next generation. Darwin gave the name natural selection to this process and he assumed that this gradual change in the characteristics of a population could give rise to a new species in certain circumstances. One circumstance which is seen as one of the most likely for the creation of a new species is when one part of a population becomes geographically isolated from the rest. If the environments in which these two parts of the original population live become significantly different then natural selection will act separately on the two parts and they will evolve in different directions. Eventually the two parts will become so different that they will no longer be compatible for sexual reproduction and hence a separate species is born.

I do hope that the reader who is new to Darwin's Theory will now be getting some idea on how the process of natural selection is a very plausible explanation of life's great variety. In Darwin's own words taken from *The Origin of Species*, "..natural selection is daily and hourly scrutinising, throughout the world, the slightest variations; rejecting those that are bad, preserving and adding up all that are good; silently and insensibly working, whenever and wherever opportunity offers, at the improvement of each organic being in relation to its conditions of life." Later on in the same chapter he re-emphasizes his point: "Natural selection acts exclusively by the preservation and accumulation of variations which are beneficial under the conditions to which each creature is exposed. The ultimate result is that each creature tends to become more and more improved in relation to its conditions."

The discovery of genes has moved the emphasis away from natural selection acting on individual organisms to natural selection acting on individual genes. I think a case can be made to show that natural selection can act not only on genes and organisms but also on species as well. It is understandable that some biologists insist that selection acts only at the level of the individual gene, after all it is the lowest common denominator; selection at the higher levels of organism or species is inevitably also acting on the genes contained within these higher levels.

During the process of sexual reproduction, genes are constantly being mixed up and redistributed among offspring and the modern version of Darwin's Theory sees natural selection as acting on all the genes shared by a sexually reproducing species. This collection of genes is known as the "gene pool" and natural selection will act on the gene pool to reduce or increase the quantities of particular genes thereby changing the average characteristics within a population. It's really a numbers game. Characteristics don't necessarily disappear, they just become more or less frequent.

We can consider a theoretical example of natural selection in action showing how a characteristic with which we are all familiar could change as a result of environmental pressure. Let's take the human species and skin colour as the characteristic which we want to consider. People who have lived for thousands of years near the equator tend to have darker skins than those who live in colder, less sunny climates. The reason for this is that the dark pigment in the skin is better able to absorb the sun's ultraviolet rays which can be dangerous and unhealthy in large quantities; the dark pigment absorbs the harmful rays thereby preventing them from entering the body. On the other hand too little ultraviolet is also unhealthy and a lack of it can cause such conditions as rickets. Therefore people who have lived for thousands of years in colder climates tend to have skin which lacks the dark pigment thereby enabling their bodies to absorb more of the weak ultraviolet rays in order to keep healthy.

If, due to some astrophysical phenomenon, the earth tilted on its axis and fair skinned people found themselves on the equator there would be an environmental pressure i.e. too much ultraviolet, acting on these people which should allow natural selection to spring into action. In this case, natural selection would manifest itself with an increase in the number of cases of fatal sunstroke and a more frequent incidence of skin cancer and people with exceptionally light skin would be more vulnerable than those with slightly darker skin. Of course most people would survive but even a slight imbalance in an individual's chance of survival could have a cumulative effect over thousands of generations and result in a tendency for darker skinned people to survive. More technically, genes for darker skin would tend to become more numerous in the gene pool than genes for lighter skin. Changes in a characteristic such as skin colour could take tens of thousands of years but as life on earth appeared at least 3500 million years ago then there has been plenty of time for gradual changes to be effective.

According to twentieth century Darwinists, random mutations of genes *are* the source of variation in a population and in our example any mutation which happened to give an offspring a darker skin would tend to be retained in the gene pool because of the improved survival chances of the offspring. Most mutations, because they occur on a random basis, will not be advantageous and therefore will not be retained in the gene pool but favourable mutations are retained and hence evolution moves in a direction which tends to make an organism better equipped to survive in its environment. It is this constant selection of useful characteristics which allows evolution to proceed. According to Darwinism, the mechanism of evolution is natural selection and the raw material of evolution is the variety caused by random mutations. Natural selection acts on the raw material to produce change in a direction which tends to make organisms remarkably at home in their particular "niche."

This description of Darwin's Theory of Evolution has ignored many twentieth century arguments concerning the finer details but the main principles are the same now as they were in 1859. In summary the theory says that evolution occurs because pressure from the environment can change the characteristics of a population by acting on the variety produced by random mutations to favour those genes and their related characteristics which are better adapted to the prevailing conditions. The definition of environment is very broad ranging; it includes temperature, light, wind, gravity, terrain, competition from other species for food, action by predators, competition between members of the same species for food, territory and mates, disease and of course the composition of the air and any other medium in which the organism lives.

Darwin himself summarised the theory in the following words: "As many more individuals of each species are born than can possibly survive; and as, consequently, there is a frequently recurring struggle for existence, it follows that any being, if it vary however slightly in a manner profitable to itself, under the complex and sometimes varying conditions of life, will have a better chance of surviving and thus be naturally selected. From the strong principles of inheritance, any selected variety will tend to propagate its new and modified form."

The main evidence that we have for Darwin's Theory is that organisms of all types appear to be very well adapted to their environments. The theory of natural selection is a simple and satisfactory explanation of this phenomenon. Darwin's book *The Origin of Species* is a meticulous catalogue of natural phenomena which he used as evidence for his theory and in particular his observation of bird life on the

Galapagos islands gave him his best evidence to support the new theory.

I am not convinced that Darwin was as obsessed with the correctness of his theory as are his twentieth century followers. It is said that he delayed publishing his theory for twenty years because of the effect that his ideas would have on the religious establishment of his day. In fact, Darwin was a competitive scientist; whatever his reasons for delaying publication he soon put them aside when the same theory was being proposed by another meticulous observer of nature in the wild, Alfred Wallace. The majority of objections raised against the new theory were rational scientific criticisms and were not an hysterical religious outcry as is popularly portrayed. It is more likely that he delayed publishing because he wanted to collect more and more evidence for the theory, evidence which he eventually included in his book *The Origin of Species*. This collection of facts not only would help convince the reader but it also helped to convince himself that the theory was correct. Most twentieth century writers have forgotten that Darwin constantly stressed that he did not think that natural selection was the only driving force of evolution. "I am convinced that natural selection has been the main but not the exclusive means of modification." He knew in his heart that it was difficult to understand how natural selection could create the more complex biological structures which we observe in nature.

This book will not attempt to belittle the impact that Charles Darwin has had on scientific thought. He is one of those rare original thinkers for whom the epithet "genius" is suitable. There can be no scientific word in everyday use more often pronounced by scientists and laymen alike than "Darwin." He was a giant among scientists and was responsible for opening a vigorous debate which has lasted to this day.

Leading evolutionists of today place different emphasis on the various facets of evolutionary theory. Some argue that evolution is not smooth but proceeds in jumps; others say that natural selection acts solely on individual genes and not on individual organisms or species. There are arguments amongst taxonomists as to how to determine relationships between species. Non-biologists are working independently on theories that prove that evolution is a natural consequence of the laws governing order and chaos in the universe. Micro-biologists who are unravelling the chemistry of genes and embryonic development have their own innumerable ideas and theories. I would say that the study of evolution is now so broad that it is becoming increasingly difficult for an individual scientist to become familiar with all the academic fields which have a bearing on the subject. I might even dare to suggest that the mass of detail within the different biological disciplines is making the overall picture somewhat cloudy and confused. This book is an attempt to get back to basics and to simplify the subject by discussing only those points which are of major importance.

There are even arguments as to what the subject matter is. Some argue strongly against any attempt to place the human species at the top of an hierarchical tree. Evolution, they say does not necessarily lead to more complexity but new species can be more or less complex depending on the particular pressures exerted by the changing environment. There is certainly no hierarchy in the basic machinery of genes and cell production; similar mechanisms using the same chemical language occur in both simple organisms and human beings. It is argued that an erroneous human-biased view of the natural world could result from placing ourselves at the top of a tree of ever increasing complexity. Despite these precautionary warnings concerning the folly of self-centredness I believe that the man in the street, the layman to whom this book is directed, wants to know above all else how we get from simple organisms to humans. He wants to understand something more about himself, his origins and his future as a species and this book will be primarily concerned with exploring the mechanism which led from those first simple organisms to the human species. There are three important prerequisites which we must accept as fact before it is worthwhile exploring possible mechanisms of evolution. The first prerequisite is the acceptance that evolution has indeed happened at all. This book is concerned with the mechanism of evolution and it takes for granted the fact that evolution has occurred. The fossil record which shows new species appearing and old species disappearing over millions of years is accepted as evidence for evolution. The only alternative explanation for this fossil evidence is that we are being misled and one species is not evolving from another but new species are appearing spontaneously by another unknown mechanism. Well, I believe in neither magic nor Divine Creation and this book is concerned not with justifying evolution but with explaining the phenomenon of evolution which is assumed to have occurred.

The second prerequisite concerns the question of how it all began. The fossil record shows us that primitive organisms were active on earth more than 3500 million years ago. The earth itself is 4600 million years old and it would have been hundreds of millions of years before the surface of the earth was cool enough for life to be possible. So it seems that life was formed fairly soon after the conditions existed which could support life. You would think therefore that the jump from non-life to life would be an event which happens fairly readily and it should be an event which we can understand and perhaps reproduce in the laboratory. Well, unfortunately this is not the case. The primitive organisms which appeared on earth so quickly after the earth cooled are incredibly complex chemicals and are made of similar building blocks to those constructing organisms such as humans. There appears to be no halfway house between inert non-life chemicals and the complex chemical factory which is essential for life. It is difficult to understate the complexity of a living cell and the following quote from Michael Denton's stimulating book *Evolution: A Theory in Crisis* will serve to illustrate this complexity: "To grasp the reality of life as it has been revealed by molecular biology, we must magnify a cell a thousand million times until it is twenty kilometres in diameter and resembles a giant airship large enough to cover a great city like London or New York. What we would then see would be an object of unparalleled complexity and adaptive design. On the surface of the cell we would see millions of openings, like the port holes of a vast ship, opening and closing to allow a continual stream of materials to flow in and out. If we were to enter one of these openings we would find ourselves in a world of supreme technology and bewildering complexity. We would see endless highly organised corridors and conduits branching in every direction away from the perimeter of the cell, some leading to the central memory bank in the nucleus and others to assembly plants and processing units. The nucleus itself would be a vast spherical chamber more than a kilometre in diameter, resembling a geodesic dome inside of which we would see, all neatly stacked together in ordered arrays, the miles of coiled chains of the DNA molecules. A huge range of products and raw materials would shuttle along all the manifold conduits in a highly ordered fashion to and from all the various assembly plants in the outer regions of the cell.

We would wonder at the level of control implicit in the movement of so many objects down so many seemingly endless conduits, all in perfect unison. We would see all around us, in every direction we looked, all sorts of robot-like machines. We would notice that the simplest of the functional components of the cell, the protein molecules, were astonishingly complex pieces of molecular machinery, each one consisting of about three thousand atoms arranged in highly organised 3-D spatial conformation. We would wonder even more as we watched the strangely purposeful activities of these weird molecular machines, particularly when we realised that, despite all our accumulated knowledge of physics and chemistry, the task of designing one such molecular machine - that is one single functional protein molecule - would be completely beyond our capacity at present and will probably not be achieved until at least the beginning of the next century. Yet the life of the cell depends on the integrated activities of thousands, certainly tens, and probably hundreds of thousands of different protein molecules."

This elegant quotation says it all. How can we be expected to give any credibility to a theory which does not adequately explain the origins of these complex chemical factories. Most of this book concerns the question of how we get from the simple organisms which first appeared on earth more than 3500 million years ago, to the more sophisticated organisms which exist today. For the time being we shall skip over the enigma of life's origins and instead concentrate on what happens after that. I will return to the problem of the origin of life in the final chapter but in principle let us accept that the first primitive forms of life arose by chance from simple non-life chemicals. Without this assumption, the only alternative explanation is that the first molecules of life were created by a method which is totally beyond our current knowledge or understanding. If we have no concepts to explain life's origins then it is hardly worth trying to understand the various stages that life passed through since its beginning. If the original spark of life did not begin in a way which is explainable (albeit an unlikely explanation) in terms of our current knowledge of chemistry and physics then an understanding of its structure, purpose and evolution will not be available to us using our limited knowledge.

The third prerequisite is related to the previously discussed problem of life's origins but it concerns not just the chemistry of the original molecule of life but the chemistry of all existing life processes. We have to accept that all the apparently fantastic organic processes from sexual reproduction and embryonic development to the workings of a fully grown organism can be analysed as being the end result of a series of understandable chemical reactions. Again, if life processes involved some kind of power which is beyond our current awareness then the study of evolution using current scientific knowledge would be futile. In fact we are constantly proving that this prerequisite holds good by meticulously analysing complicated organic processes and turning them into understandable chemical reactions. These chemical reactions are predictable given the conditions in which they take place. There is nothing magical about the development of an embryo or the workings of the human body, it is just that the chemical reactions behind it all are incredibly difficult to analyse.

Now that we have set the ground rules for our study of evolution we can have a closer look at the subject matter itself. This book is concerned with discovering a plausible explanation of the mechanism which allows simple organisms to evolve into human beings and it is now a good point to identify the important steps that evolution has taken en route.

The first simple organism which appeared on earth would have had little or no ability to react to the environment. The secret of its success was merely its ability to make copies of itself. If it somehow managed to spread outside the environment in which it was created then it would either not survive in its new surroundings or, according to modern Darwinism, it might undergo a chance chemical change, a random mutation of its genes, to enable it to adapt to its new environment and survive. Any change in the characteristics of the early organisms appeared to rely totally on genetic mutations alone. Once the organism was formed, its die was cast and it had no ability to react to the environment during its own lifetime.

Eventually however a mutation occurred which greatly improved the ability of the organism to cope with different environments and the primitive organism was given the ability to react to the environment during its own lifespan. This ability to react to the environment in the short term was far more efficient as a survival mechanism than waiting for genetic mutations to alter a characteristic. Organisms which achieved this improved survival mechanism would be better able to survive in a variable environment than their more primitive cousins. This short term flexibility might have taken the form of an ability to develop an extra coating of protective "skin" in response to a drop in temperature or maybe a simple ability to remove itself from the intense heat of the sun. More sophisticated flexibility is seen where plants can take on different appearances depending on the environment in which they are sown. These abilities would have been auto-responses like our own knee-jerk; there would be no choice in the matter, just an automatic chemical response to a particular stimulus. Genes for useful automatic responses were retained and more and more automatic responses were accumulated. Other parts of the organism which had not gained the ability to instantly respond to an external stimulus would continue to evolve in a direction which made them more suited to their environment (or so Darwin tells us).

When considering the end result of evolution it is almost as if the aim of the exercise is to improve the ability of the organism to survive and obviously those organisms which are good at this pursuit are the ones which *do* survive and thereby pass on their genes to the next generation. Or to put it another way, good survivors survive. This statement has encouraged philosophers and students of logic to attack Darwinism on the grounds that it is meaningless. The statement should be re-phrased to read "organisms which have developed particular chemical and physiological characteristics will survive in certain environments."

There does indeed appear to be an unconscious aim of evolution. This unconscious aim is to improve the survival chances of the organism and its future descendants.

Evolution, in fact, does not know where it is going; there is no conscious effort pushing evolution in a certain direction. We get the illusion of direction because it is only the good survivors who are around today.

Darwin tells us that improvements come about in response to the environment and that these improvements are accumulated over time. In the days before sexual reproduction an improvement due to a favourable mutation would spread into the population by asexual reproduction. Asexual reproduction produces offspring that are almost an identical copy of the parent. The first organism to have a favourable mutation in its genes could pass on that improvement to its direct descendants. Offspring of contemporaries of that lucky first organism could never include this favourable mutation in their collection of genes (or genome as it is technically called) unless by some rare chance the same mutation occurred again. With the arrival of the mechanism of sexual reproduction, which itself was the result of a genetic mutation, favourable mutations were now available to all future generations, not just the descendants of the first carrier of the mutation. This is because the consequence of sexual reproduction is that the offspring inherits genes from two parents, 50% from each, and therefore any offspring in the population has a chance of inheriting the favourable mutation. In this way numerous favourable mutations could accumulate in the same individual and "super organisms" could evolve which would not only have numerous strategies for survival but would also pass on this collection of useful genes to the next generation. Thus we have another example the unconscious aim of the exercise in action; evolution has come up with a much improved mechanism for survival.

It was necessary at this stage to introduce the concept of sexual reproduction as it makes the idea of the gene pool a little clearer and it is a necessary requirement for discussing another significant stage in evolution. This stage is again an improvement in the ability of an organism or a species to survive. The variety which is so essential for natural selection is contained within the gene pool. Some characteristics of an organism are so little favoured by the prevailing conditions that their frequency in the gene pool is low but nevertheless they are still available to natural selection when the conditions change to make use of them. I consider this store of variety to be a medium term strategy for survival. The organism has hidden within its gene pool the ability to change its characteristics over a number of generations in response to a rapid change in environment. The gene pool is slowly added to by random mutations and it is these random mutations which could be referred to as a long term strategy for survival. Microbiologists recognise a strategy which falls somewhere between the long and medium term wherein a particular mutation occurs on a regular and predictable basis every so many thousand generations. As with random mutations in general these regularly occurring mutations will not always be advantageous, some if not most, will be harmful to the organism. Nevertheless it is the occurrence and store of mutations which gives the species the ability to adapt to a changing or variable environment. This improved mechanism for survival is a further example of the unconscious aim of evolution, the gradual searching out and discovery of a more efficient method of coping with the outside world. The total gene pool can hold far more variety than any individual organism and this stored arsenal of survival weapons is supplemented even more by random and regularly occurring mutations.

These survival mechanisms are passed down to future generations in a perfectly reliable manner and some of these inherited characteristics include the short term survival mechanisms which we have already met; mechanisms which give the organism the ability to react instantly or at least within its own lifetime to an external stimulus. Examples of this short term mechanism in humans are our ability to develop bigger muscles in response to exercise or the thickening of the skin on the hand in response to manual labour, or the darkening of the skin in response to ultraviolet light.

These short term mechanisms are fast acting survival strategies when compared to the slow accumulation of new genes or the lengthy period of time needed for low frequency genes to become more prevalent in the gene pool. The organism however is given no choice in its reaction to a stimulus in that these reactions are purely automatic responses. The next stage in evolution gave the organism the ability to chose between alternative responses to a stimulus. The development of a nervous system and primitive brain allowed the organism to choose between alternatives on the basis of stored information. With the ability to perform complex tasks such as hunting, storing food and building shelters the organism greatly improved its chances of survival. Evolution had yet again come up with an improved survival mechanism.

Not all complex behaviour in organisms is governed by choice, much of the behaviour patterns which we observe in the lower animals is rigidly inherited via the genes. This fixed type of behaviour is merely a more sophisticated version of the automatic responses which we have already met and this genetically transmitted behaviour is performed in exactly the same way by succeeding generations. Evolution then gradually came across a much more efficient method of performing tasks. Instead of a specific ability being programmed into the genes an ability to perform a multitude of various tasks was developed; a brain which was capable of choosing between alternatives; a brain which gave the organism a task performing ability which was not pre-programmed by the genes.

It is intriguing to attempt to distinguish between the type of behaviour which we might call instinctive and the type of behaviour which is considered to be learned. The distinction is difficult to define because the transition between the various stages of evolution is a gradual affair with each stage co-existing with the next stage and indeed all the stages continuing to exist and evolve even in the presence of superior stages. There is no distinct dividing line between these arbitrary stages and this makes the task of distinguishing instinctive behaviour from learned behaviour somewhat difficult. As the brain evolved, certain types of behaviour which were once controlled by genes could now be transmitted from one generation to another in a social way. If this method of learned behaviour could pass on a skill from one generation to the next in a way that was as reliable as the genetically transmitted method then evolution would favour this more efficient and flexible method. It would be extremely difficult for more complex behaviour patterns to be transmitted genetically because the genes would have to give the brain an incredible amount of information about the kind of world the organism was being born into. The ability to pass on skills in a learned or social way was another important stage in evolution; as with all the other stages, it gave the organism or the species an important survival advantage.

The limit of complex behaviour which is possible with little or no element of learning can be seen in the fascinating world of the social insects. Honey bees for example have an incredibly wide range of behaviour patterns which is thought to be almost entirely instinctive. Not only do bees live in a strictly ordered society with each type of bee assigned specific duties, but they also have evolved elaborate ways of communicating with each other. Their means of communication include the performance of a dance routine to inform others of the direction, distance and amounts of a food source and they can pass on orders down the ranks by the secretion of various scents and odours. At the other end of the scale, the human species has hardly any instinctive social behaviour at all, nearly all our behaviour is learned.

In our search for the various stages of evolution we are

now approaching ourselves, the human species. It is very difficult to say precisely what advantage evolution has given us. Not only are we great survivors in the earth's environment but we are so skilful that we are even capable of controlling the environment for our own ends. Primitive animals are capable of passing on behaviour patterns in a social way but humans can rapidly improve on socially inherited skills and this ever increasing store of accumulated knowledge is passed down from one generation to the next.

Some scientists and philosophers say that language marks us out as different from other animals and I must admit that it would be difficult for us to progress in this accelerating manner without language. My own feeling is that language is somehow inextricably connected with our ability to reason, plan, think and store knowledge. I like to think that our advantage over other species was the evolution of a better brain and not merely the evolution of a brain related skill such as language.

Let us now summarize the stages of evolution which have been outlined above and place them in order of improving survival mechanism.

1. Survival dependent on random mutations.

2. Evolution of simple ability to react automatically to external stimulus.

3. Sexual reproduction and a store of potential variety within the gene pool with the added advantage of regularly occurring mutations.

4. Genes which induce automatic complex behaviour in the organism.

5. Genes which allow multiple choice, non-automatic behaviour.

6. A brain sufficiently developed to allow accumulation of knowledge stored non-genetically and passed down from generation to generation.

The six stages are purely arbitrary and are not meant to be mutually exclusive. Each stage represents an improved survival mechanism but the appearance of a new stage did not mean the extinction of an inferior stage; evolution only rejects a phenomenon if it is actually harmful or positively disadvantageous; one survival mechanism is still advantageous even if it is inferior to another. The main purpose in identifying and listing these six stages is to enable reference to be made to them in the following chapters.

Chapter Two

Two Major Weaknesses with Darwin's Theory

You would have to be an extremely dull human being not to have questioned the reason for your existence. The study of evolution is really a scientific approach to solving this mystery of mysteries. I am often surprised that more people are not interested in evolution. Perhaps most people think that the mystery of life has already been explained by Darwin or maybe people get so confused and bogged down with the subject that they give up trying to understand it. Then there are those who reject the fact of evolution and prefer to believe in Divine Creation instead. I cannot be too critical of believers in Darwin because I was once an ardent supporter of his theory but I am now convinced that Darwin's Theory is wrong; natural selection is not the main mechanism of evolution and this book is an attempt to argue the case against Darwin's Theory.

A new theory has not yet been mentioned or even hinted at and before you are introduced to it I want to critically examine two fundamental requirements of Darwin's Theory and thereby hope to persuade the reader that they are by no means established facts and are far from convincing.

The two requirements are 1. Random mutations in the gene pool can provide the fuel on which natural selection acts and 2. The environment is the driving force of evolution using natural selection as its mechanism.

For the sake of lucidity and with the layman in mind I do not intend to give a detailed description of the chemical composition of genes and how they are thought to trigger a series of chemical reactions which lead to the building of a finished organism. For the time being a brief description will suffice.

It was in 1953 that James Watson and Francis Crick

were able to create a model of the structure of genetic material. The chemical name for this material is deoxyribose nucleic acid or DNA for short. It looks something like a twisted ladder with the rungs of the ladder being made of four chemical bases, adenine, thymine, guanine and cytosine. It is the order in which these bases are strung along the ladder which comprises the program which directs the building of the organism. A mutation is no more than a change in the sequence of the bases and modern research has discovered some of the ways in which these mutations can take place. In the series of events which lead to sexual reproduction the sequence of the bases is rearranged and during the duplication of the genome (a crucial step in reproduction) copying errors can occur. According to Darwin's Theory these random changes in the sequence of bases are the raw material of evolution but the survival of a particular change is not a random event because it is only useful changes which will survive in the gene pool. This is the crucial argument put forward by Darwinists to convince us that an apparently random event can lead to evolutionary change. Natural selection acts as a sieve for random mutations and only those changes which are not harmful to the organism will survive; the more useful the change the better chance it has of being retained in the gene pool. Harmful changes are eliminated from the gene pool because organisms carrying the harmful mutation will probably die out before being able to pass on their genes to the next generation. Despite the sifting out process of natural selection it is still difficult to imagine how a series of initially random changes can lead to the evolution of something as complicated as a human being. The complexity of any single organ has led some biologists to doubt whether natural selection could possibly be the agent of evolution. A favourite example used to illustrate this doubt is the eye and even Darwin had this to say, "To suppose that the eye with all its inimitable contrivances for adjusting the focus to different distances, for admitting

24

different amount of light, and for the correction of spherical and chromatic aberration could have been formed by natural selection seems I freely confess, absurd in the highest degree." In fact Darwin's description of the complexity of the eye is something of an understatement; he does not mention the complexity of those parts of the brain which do most of the work when it comes to light perception. Even with the use of sophisticated computers and stereoscopic cameras it is extremely difficult to simulate just one of the many aspects of visual perception, namely our perception of distance and depth. The visual interpretation of the world by humans and other species is achieved by the interaction of a most powerful organic computer working in perfect harmony with a sophisticated lens system.

Although the natural selection route to visual perception is an unlikely event, Darwin and his followers have managed to convince us that this is how it all happens. One of the most readable defences of this unlikely mechanism is given by Richard Dawkins in his excellent book *The Blind Watchmaker* which is a modern and convincing (for some) defence of Darwinism. Dawkins argues backwards from the eye as we know it and says that you have to imagine an eye one small stage inferior to the current human eye. If the jump from one stage to the next is sufficiently small to be accounted for by a single favourable mutation then you should be able to extrapolate backwards to the point where we have no eye at all. The only limiting factor argues Dawkins would be the availability of sufficient time for all the stages to have happened. Dawkins' contention is that the time *has* been available and that given as many as a hundred million stages in the development of the eye then "we should be able to construct a plausible series of tiny graduations linking a human eye to just about anything!" Yes, I agree, there has been a lot of time available and unless you believe in Divine Creation then there must be a finite number of steps linking no eye to the human eye but the crucial question is, how probable is even *one* favourable step towards the human eye given the apparently millions of possible mutations. The end result of the hundred million mutations is a wonderfully coordinated system of muscles, lenses, bone structures and computing and therefore each and every mutation must in some small way be a coordinated move in the right direction. The finished eye is a masterpiece of complexity and it must follow that each and every other stage in its evolution was also a highly complex, fully integrated system. I find it difficult to imagine even *one* step in the right direction never mind the hundred million which might be required. The sticking point is the concept of random mutations and we must ask ourselves whether a truly random system can be expected to come up with complex improvements to an already complex mechanism. I have not forgotten the sifting out process of natural selection but I am arguing against the possibility of a useful mutation happening in the first place.

Attempts to quantify the probabilities involved in creating the chemicals of life by chance end up by proving that such random events are all but impossible. There are simply too many possibilities to choose from and the chance of achieving even one correct solution on a random basis is most unlikely and achieving millions of correct solutions one after another is quite unbelievable. To get some idea of perspective into what can be achieved by chance let us consider that age old means of settling a dispute, tossing a coin. We all know that the chances of getting a head as opposed to a tail is 50% certain and to get two heads in succession is 25% certain but how long do you think it would take to come up with a sequence of say 50 consecutive heads? Well don't put any money on it because if you started tossing coins tomorrow it would probably take more than 10,000 million years to achieve a run of 50 consecutive heads.

Mathematicians are surprised that biologists so readily

accept that a random process can supply the raw material on which natural selection can act and in 1967 a conference was held entitled "Mathematical Challenges of the Neo-Darwinian Theory of Evolution" to discuss the problem. The fact that the objections raised by mathematicians have been largely ignored demonstrates either a reluctance to accept such a devastating blow to Darwinism or reflects the sheer arrogance of those biologists who study evolution.

Michael Denton in his book *Evolution: A Theory in Crisis* is severe in his criticism of the reliance on random mutations when he says "To the sceptic, the proposition that the genetic programs of higher organisms, consisting of something close to a thousand million bits of information, equivalent to the sequence of letters in a small library of one thousand volumes, containing in encoded form countless thousands of intricate algorithms controlling, specifying and ordering the growth and development of billions and billions of cells into the form of a complex organism, were composed by a purely random process is simply an affront to reason. But to the Darwinist the idea is accepted without a ripple of doubt."

The probability of creating the eye as a result of a random process is about as likely as tossing ten tons of scrap metal in the air and hoping it comes down as an aero-engine. No, Mr.Dawkins, you would not get an aero-engine in a thousand years, not in a million years, not in the total age of the universe.

The staggering variety of species which currently exist (probably several million) is dwarfed by the even more staggering variety of species that has existed in the past. But an even greater quantity of variety exists not between species but within the species themselves. Nearly all members of the human species, for example, are different from each other. Taking into account the variety that exists within each species then we are talking about not just billions but trillions (thousands of billions) of different fully coordinated working bodies or phenotypes as they are technically described. (The phenotype is the body which houses the genes; it is the protective structure which allows the genes to be transported from one generation to the next). And modern Darwinists expect us to believe that each phenotype is different because of a random alteration to the program that builds and controls them.

Darwin's Theory depends on the availability of variation in the population and the source of this variation is assumed to be random mutations. We know that variation comes about by genetic rearrangement but the crucial question is, how does this rearrangement come up with useful evolutionary change or to put it another way, how does it avoid the havoc which would inevitably be created by a random shuffling of the program. I stated in chapter one that we would only consider the crucial aspects of evolutionary theory and we have now met the weak link in the whole edifice. If there is one indisputable objection to Darwin's Theory of Evolution it is the lack of a valid explanation for the existence of the raw material of evolution. Modern versions of Darwin's Theory do not explain the mechanism or system which leads to the variation which we know to exist within every species. The reason for the existence of the variation which eventually leads to new species and other evolutionary events has not yet been convincingly explained by neo-Darwinism. The source of that variation is as mysterious today as it was to Darwin and modern biologists should take a lead from *The Origin of Species* and admit, as did Darwin, that this source of variation remains unknown.

The reader may now have gathered that I find something wrong with this particular aspect of Darwin's Theory and in chapter three I will introduce you to a more likely explanation of evolution.

Let us now move on to the second requirement of Darwin's Theory and see if it fares any better.

It was Darwin's observations of nature which led him to his theory of natural selection. He observed the way in which living things are so well adapted to their environments. His explanation of this phenomenon was that natural selection was the mechanism which allows the environment to drive evolution and the nearest he could come to demonstrating his theory experimentally was to study the effects of selective breeding on domestic plants and animals. In the case of dogs we have succeeded in breeding varieties as different as a chihuahua and a St. Bernard from the original primeval wolf-like wild dog. In the dog breeding world the environment is the conscious effort of humans to develop dogs with particular characteristics; characteristics which would be useful when performing a specialized skill such as hunting or winning at Cruft's dog show. The successful breeding of different types of dog is, according to Darwinists, evolution in action albeit by *un*natural selection.

When looking for evidence of evolution in the natural world the most often quoted example of natural selection in action is the case of the peppered moth which changed its colour from white to black in the aftermath of the industrial revolution. The study of the peppered moth is regarded as a truly classical piece of field research and no good book on evolution excludes it. This volume is not necessarily a good book on evolution but I am going to tell you about the peppered moth anyway!

In the days before the industrial revolution, peppered moths were nearly all of white colouration and when they settled on the white coloured lichen which covered the trees they were well and truly camouflaged. This camouflage was crucial to their survival as the moths are a tasty part of the diet of birds. However, the industrial revolution brought with it pollution which killed off the lichen and left the dark coloured bark exposed thus neutralising the camouflage advantage of the white moths. Fortunately the original moths were not all white and there was the odd black one in the population and these black ones now had the advantage of camouflage over the white ones. Natural selection was now able to swing into action and black moths survived at a much higher rate than white moths and therefore genes for black coloration were passed onto future generations at an ever increasing rate at the expense of genes for white coloration which were rapidly reduced in the gene pool. The technical name for these moths is Biston Betularia but they are commonly known as Kettlewell's moths in deference to the biologist who carried out the original field research.

This classic study of natural selection in action has all the ingredients for academic discussion - change of environment, mutant black moths, struggle for existence, survival of the fittest - what a marvellous vindication of Darwin's Theory of Evolution. But is it? If you return to the end of chapter one and the six stages of evolution, the peppered moth phenomenon is an example of stage three; the peppered moth is drawing on its reserve of hidden variety; the hidden variety which gives the organism the ability to react to an external stimulus in the medium term. The peppered moth is drawing on its store of low frequency genes and regularly occurring mutants which only come into play when needed. The change from white to black in peppered moths was not an example of evolution, it was an example of a survival mechanism which evolution has created. In the same way, developing bigger muscles as a result of manual labour is not an example of evolution but merely an example of a short term survival mechanism conferred by evolution. The study of the peppered moth does not help us understand the development from simple organisms to human beings; the change from white to black demonstrates no advancement at all, it is really just a sidewards movement. In fact if the white lichen returned then the moths would revert to their white colour and back to square one.

This brings us back to the old argument of what we are trying to explain by studying evolution. I can only repeat that I am mainly concerned with explaining how we get from stages one to six as listed in chapter one; interesting though it is, the ability of peppered moths to change from black to white and back again gives us no definite clues in our quest to understand evolution. If the hierarchy from simple organisms to humans can be regarded as vertical then the example of the peppered moth is an horizontal phenomenon, half way up the vertical ladder.

Similarly, I am convinced that the vast majority of genes in the present canine gene pool were present in the gene pool of the ancestral wild dog but they remained dormant until brought out by intensive selective breeding. I do not believe that the great variety of dogs which now exist are the result of numerous random mutations which would be necessary if this was an example of Darwinian evolution in action. New breeds of dog were created by calling on the hidden variety of stage three. Despite all our efforts of selective breeding with dogs we still have not managed to produce a new species; even a chihuahua is capable of mating with a St.Bernard, albeit with great difficulty!

Of course, biologists will still jump up and down saying that peppered moths and domestic dogs are a fine example of the power of the environment. I would agree that changes in the environments of dogs and moths have had a dramatic effect on their characteristics but natural selection is only acting within stage three, it does not help to explain how we get from stages one to six which is *my* definition of evolution.

To vindicate Darwin's Theory, the environment must be capable of taking evolution from stage one to stage six. But does the environment really have the power to achieve this? Is the struggle for existence so intense that organisms as complex as ourselves have had to be evolved in order to survive. It is extremely appealing to believe that the environment *is* responsible for evolution but can we really accept this idea? The relationship between environment and organism is evidence for Darwin's Theory but it is extremely difficult to reconcile the minor adjustments which we know for a fact that an organism can make in response to a change in environment with the major differences between all the extremes of life on earth.

Biologists can nearly always justify a characteristic in terms of its usefulness to the organism in the environment in which the organism lives but is this fit between organism and environment merely a superficial phenomenon? Are we somehow being misled by looking only at characteristics which can be explained by environmental usefulness and ignoring the vast majority of characteristics which are unrelated to a particular environment. After all most creatures can survive in a variety of different environments and we know of many examples of animals being transported across the world by emigrating humans only to become immediately established in a foreign environment. Pigeons were never evolved to live in Trafalgar Square but it is certainly one of their favourite habitats. It would not take long for the large African carnivores to adapt, via stage three, to the cold conditions of Scotland if we made the effort to introduce them there. In fact most creatures are faced with living in an ever changing environment as the seasons change from winter to summer and as the day changes to night. What exactly are we all adapting to?

Even Darwinian biologists are sometimes puzzled when a creature migrates of its own free will and for no apparent reason into an environment to which it is less suited. Explanations of this phenomenon are extremely difficult in Darwinian terms. Darwin himself did not overstress the link between environment and organism even though it was crucial to his theory. When discussing acclimatization in *The Origin of Species* he says "The degree of adaptation of species to the climates under which they live is often overrated, and a number of plants and animals brought from different countries are here perfectly healthy. In regard to animals, several instances could be adduced of species having largely extended, within historic times, their range from warmer to cooler latitudes, and conversely. The rat and mouse have been transported to many parts of the world, and now have a far wider range than any other rodent; for they live under the cold climate of Faroe in the north and of the Falklands in the south, and on many an island in the torrid zones."

It is extremely difficult to define a particular environment and therefore how can a creature be said to have evolved in response to that environment? Biologists have refined their definition of environment and the word environment has now been replaced by "niche." Niche is an all embracing term which defines an organism's position in the ecosystem. It includes its habitat, source of food, and all its interactions with the environment and other creatures. The word environment is more commonly used in the restricted sense of describing geographical and meteorological features of a physical area. I will not be obsessive in the correct use of these two terms and I shall continue to use the word environment when the word niche would be more appropriate.

Getting back to the problem of defining environment or niche I have to admit that I have great difficulty putting into words the objections which I harbour to the accepted link between environment and evolution. It's a complicated subject and I am stalling for time by looking out into my garden for inspiration. I am lucky enough to have a large garden and my study looks out over a lawn and beyond to some woodland. We have observed more than twenty species of bird in the garden and we also get periodic visits from deer and fox not to mention innumerable grey squirrels. There is a kestrel hovering overhead and I hope he catches one of those mice who thinks he has a right to share my home. According to Darwin this wonderful bird has been shaped by the environment to give it the extraordinary ability to hunt in this way. Even in a swirling wind the Kestrel hovers with meticulous control which enables it to keep its eyes steady, thereby scanning the ground with its telescopic vision and preparing to swoop down on its unsuspecting prey. But how can the environment be said to have evolved such a species? Is Darwin saying that the ancestor of the Kestrel found itself trapped in a part of space fifty feet above the ground with no alternative but to develop the skills necessary to survive in that environment? Other birds have not developed such specialist skills and they seem to cope OK in a similar environment so why has the Kestrel taken that evolutionary pathway? If the unconscious aim of evolution is an improved ability to survive then why has the Kestrel come up with that particular survival mechanism out of an almost unlimited choice of alternatives. With all those complicated skills the evolutionary route of the Kestrel was hardly an easy option. Surely evolution will attempt to solve survival problems in the easiest way possible and not discover an unnecessarily complicated solution. In fact if you apply this thinking to the whole of life on earth then why have we evolved any complex features at all; why didn't life remain microscopic in size, develop an impenetrable skin and evolve the ability to hibernate for 99.9% of the time only waking up occasionally for food. Surely that would have been a perfect survival strategy coping brilliantly with the environment; why go for something so complicated and specialized that the organism remains highly vulnerable to changes in the environment.

I think that we would all agree that the Kestrel appears to live in a niche to which it is well suited but the question I have often asked myself is did the Kestrel develop its skills, before it found its niche or did it, as Darwin's Theory would suggest, arrive in the niche first and develop its skills in responses to its new situation?

The Kestrel strikes me as being an example of unnecessary complication given that the aim is survival. How many other examples of unnecessary development can we find? What about the human brain? What powerful environmental factor could possibly have necessitated the evolution of an organic computer more powerful than anything we can create with our advanced technology? The evolution of the brain occurred long before we were able to make full use of such a powerful tool; as time goes on we seem to make more and more use of something that was not initially designed for such a complicated world. Is it just a fluke of evolution that something shaped by an environment of hunting and gathering is now being put to good use exploring the mysteries of the universe? I see very little evidence of environmental pressures so powerful to necessitate the evolution of an advanced brain and if these pressures existed why are we the only species to have been subject to them?

Having argued strongly against the power of the environment to shape life on earth I fully accept the part that natural selection plays in fine tuning the organism to the environment but only within the narrow bounds of stage three. Stage three gives the organism an excellent medium term mechanism to adapt to a local environment but I believe that these adaptations are superficial and not crucial to the evolutionary survival of the organism. These stage three adaptations are not examples of evolution in action; our definition of evolution is the journey from stage one to stage six and should not be confused with the calling on of survival strategies within stage three. It is the phenomena of stage three which mistakenly led Darwin to believe that natural selection could explain the broad spectrum of evolution. I see very little evidence that the environment is the powerful sculptor of life that it is meant to be but there *are* innumerable examples of species drawing on stage three reserves to improve their situation.

The arguments against the link between organism and environment are almost philosophical and I feel inadequate to make a better case for my doubts. I invite the readers to consider the evidence which is all around us on earth and ask themselves if they really believe that environmental pressures could really be responsible for creating the millions of different species which currently exist?

Not all biologists are equally obsessive in their enthusiasm to explain every characteristic in terms of a response to the environment. Darwin himself, in characteristically modest style, said, "we may easily err in attributing importance to characters, and in believing that they have been developed through natural selection. We must by no means overlook the effects of spontaneous variations."

There are now an influential body of modern biologists who accept the more moderate view that certain characteristics are not linked to environmental pressure but which exist merely because of restrictions placed on development by the laws of chemistry and physics. In other words a lot of what we see in nature is merely the result of a collection of chemicals taking on a form which is governed by the physical properties of the molecules involved. Evolution will sometimes find these incidental properties of chemicals advantageous but more often than not the finished product will be a less than perfect compromise between what is ideally required and what is possible.

Despite the body of opinion which now recognises the part played by the physical properties of chemicals there remains an obsessive desire amongst biologists to explain almost every conceivable characteristic in terms of its adaptive significance. I am often amazed at the ingenuity of the explanations of the usefulness of a particular characteristic and how that characteristic has evolved. It is always important to account for the various stages in the evolution of a particular characteristic and each stage must be shown to confer some advantage to the organism. Occasionally, even the most imaginative biologist has difficulty in explaining how a particular feature could have developed by the Darwinian rule of environmental adaptation. Sometimes they are forced reluctantly to the conclusion that the characteristic was not initially developed for the use to which it was eventually put and the expression "pre-adaptation" has been coined to take account of these situations. The classical example of pre-adaptation is the evolution of the bird's wing. It was difficult to find a use for a half evolved wing; it certainly would not have been any good for flying so why did it continue to evolve down the route which eventually led to the aero-dynamically efficient structure which allowed reptiles to become airborne? Biologists have been forced to the conclusion that the world's first primitive wings were evolved to keep the reptile-like creatures warm. Then suddenly by accident the reptile-like creature found a second use for these insulating feathers and it was away, flying.

In fact, feathers were only one of many adaptations needed to make the creatures airborne. The tail disappeared, the skull and bones became lighter, the centre of gravity changed and an improved brain along with keener eyesight and an altered metabolism were needed to cope with the developing aero-dynamic skills.

It does not require very much of a change of attitude away from trying to justify everything in terms of adaptation to realise that pre-adaptation is a more sensible explanation of almost every significant step taken by evolution. For example, the first fish to venture onto land must have been in some way able to survive in air before it could take its first tentative steps; after all, dead fish would not have been much use to natural selection.

The recognition by biologists of the laws relating to form and the role that pre-adaptation could play in evolution is a sign that it is becoming increasingly difficult to stick to the rigid doctrine of Darwinism. I shall talk more about pre-adaptation in chapter three after having introduced you to the new theory.

Another example of the difficulty which Darwinism faces when trying to explain a particular aspect of evolution concerns the tendency of many creatures to gradually increase their body size, generation by generation.

Dinosaurs are the classic example of the result of this tendency but many animals existing today, including ourselves, have evolved from smaller, primitive ancestors. The two most popular explanations of this phenomenon are firstly that the competition for mates between members of the same species gives an advantage to bigger bodies and secondly a larger body has the advantage of retaining heat more efficiently. The latter advantage is the consequence of surface area increasing at a slower rate than body bulk and as the body generates heat and surface area loses heat then a bigger body loses relatively less heat than a smaller one. No doubt if the fossil record proved that bodies tended to get smaller over evolutionary time then biologists would claim that keeping cool was an advantage in the heat of summer and when it comes to courting then small is beautiful. There is perhaps some logic in the accepted argument until you start to consider the *disadvantages* of large bodies. The whole mechanics and physiology of a large body are more cumbersome and less efficient than a small body, not to mention the increase in food and water that a large body needs. A large body makes you so much more visible to predators but with a small body you can hide in caves or holes in trees or even under a leaf. You can't even shelter from the elements if your body gets to the size of a dinosaur whereas a small creature can dig itself into the ground or build itself a nest. I cannot think of any development which is so obviously disadvantageous as a big body. This tendency towards larger bodies is another aspect of evolution which is difficult to explain in purely Darwinian terms and it demonstrates that the connection between environment and evolution is not as definite as some biologists have led us to believe.

Evolution is capable of creating the most amazing diversity, even when there is little variation in the environment. The great lakes of the African Rift valley contain hundreds of different species of fish, all sharing the same environment. It is almost as if there is a mechanism which encourages diversification as it is difficult to understand how so many different species could have been created in such a uniform environment.

The same addiction to diversity can be seen in the orchid which some botanists think exists in as many as 30,000 different species. Why does only the orchid demonstrate this amazing variety? Why not elephants or rhinos?

The best examples of diversity however, occur amongst creatures which are either highly uninteresting such as flies and insects or amongst organisms which are invisible to us such as bacteria. The Hawaiian islands which have gone their own evolutionary way for millions of years are home to thousands of species of insects which are unique to those islands. It is difficult to explain such diversity in terms of responses to different environmental situations.

If the pressure of the environment *was* responsible for all this diversity then perhaps we should expect a much better fit between organism and environment but in fact we see many examples of creatures which survive quite happily despite having characteristics which do not appear to have any adaptive advantage. For example, although the predominance of cold blooded fish indicates a certain advantage over warm blooded fish there are plenty of examples of warm blooded fish who buck the trend. There is, in other words, evidence of diversity for its own sake.

We must remember that our definition of environment is very broad ranging and biologists would point to the power of competition between species and amongst individuals of the same species as an important factor in driving evolution. In fact the much misused expression "The survival of the fittest" would suggest that most of evolution results from this type of pressure. A closer examination, however, indicates that species would prefer to develop a strategy which actually reduces competition and encourages peaceful co-habitation rather than risk death and destruction in open combat. Rules regarding territory are a good example of this avoidance of conflict and specialization of diet is another way of avoiding competition. Many creatures have evolved strategies for co-operation rather than conflict and when animals such as deer fight over territory or mates it is very seldom serious warfare but more an elaborate mock ritualistic fight without the slightest chance of injury. In other words, nature is good at neutralizing the pressures which are commonly thought to be the driving forces of evolution.

A detailed investigation of the relationship between organism and environment reveals few patterns and rules. Different species share the same environment but adapt to that environment in a variety of different ways. Thus we see creatures large and small, browsers and grazers, surface dwellers and underground dwellers, water seekers and land lubbers, specialists and generalists all sharing the same geographic area. At a different level of classification most creatures are in fact extremely versatile and unspecialized when it comes to their relationship with the environment. The specialization which we would expect from Darwin's Theory is actually discouraged by evolution because specialization often leads to extinction. The best survivors would be the jack-of-all-trades; organisms which could survive in the greatest variety of circumstances and not just in one narrow environment.

Gordon Rattray Taylor in *The Great Evolution Mystery* gives numerous excellent examples of evolution occurring without any convincing link to environmental pressure. He points out that squids and octopi developed their ability to dart at prey using a powerful system of jet propulsion well before there was anything worth darting at! He also comments on the equally puzzling fact that many creatures fail to make the most out of their environments and some hardly evolve at all.

Rather than continually try and justify the evolution of an organism in terms of relationship with its environment I find it much easier to suppose that evolution has an innumerable variety of potentially successful strategies available to it in any particular environment. After all, the vast majority of species co-habit in similar environments with other very different species and we have mentioned before that many creatures are able to survive quite happily in different environments. The contention that each species has been adapted to its particular niche is rather a weak or circular argument in that a particular niche can only be defined after the event. In other words a niche is merely a description of where a particular organism has finished up, it tells us nothing of the forces which shaped its development.

In summary I admit that there is a degree of interaction between organism and environment but this amounts to only a superficial shaping of the organism via the fine tuning of stage three. The major choices of strategy made by evolution such as whether to fly, swim, dig holes or live in trees cannot easily be explained by the exertions of the environment.

Hopefully, chapter three will introduce you to an idea which gives a more satisfactory explanation of the diversity of life.

Chapter Three

The New Theory

The great heroes of the modern age are not politicians, soldiers or film stars but they are the researchers and cameramen who dedicate their lives to producing wild-life programmes for television. They spend weeks, months, sometimes years patiently tracking, observing and recording nature in its natural habitat to enable the likes of you and me to benefit from their Herculean labours in a forty minute wildlife programme. Apart from the view of my garden from my study I remain a great television wildlife enthusiast. I have in fact been on a few African and Indian safaris and despite the excitement of the real thing there is no substitute for the excellence of a David Attenborough documentary. A week spent tiger spotting in the Indian jungle inevitably ends in disappointment but a re-run of Naresh Bedi's "Saving the Tiger" never fails to thrill. The producers of these programmes give us a marvellous insight into the habits and lifestyle of some of the wonderful creatures with whom we share this planet.

How often have you wondered at the ingenuity of nature? How often have you been mesmerised and astonished at the cleverness of it all? There appears to be no limit to the tricks which nature will conjure up in order to make an organism better equipped to survive. A variety of shapes and sizes combined with a seemingly endless repertoire of skills, strategies and behaviour endow each species with its own particular survival mode. There is hardly anything invented by man that had not previously existed in nature: light production, the ability to produce and sense electric currents, radar, computers, flight, rotary and jet propulsion, poison, high pressure and low pressure survival mechanisms, navigation using the earth's magnetic field, antennae receptors, liquid crystal etc, etc. According to Darwinian doctrine this unlimited resourcefulness is a response to environmental pressures. According to Darwinists this bewildering array of survival mechanisms is the result of natural selection acting on random mutations.

Every owner of a television set can now be aware of the true extent of the variety of life on earth. One of my favourite life forms is the worm known as the Microstomum which occasionally digests a small creature called the Hydra. The Hydra is a small underwater creature which is known for its use of stinging cells which it projects from its surface as a means of defence. Despite this effective defence mechanism, the Microstomum not only swallows the Hydra but preserves the stinging cells and proceeds to use these weapons for its own defence. The stinging cells of the Hydra are absorbed by the Microstomum but instead of digesting them, the cells pass through the body and take up positions on the surface of the Microstomum in order to protect it.

I have always been fascinated by the way in which many creatures find their way home to their original nesting site after migrating half way across the world. We all know of the salmon's incredible ability to find its way home after travelling across hundreds of miles of ocean and we have also heard of the strange story of the eels and the Sargasso Sea, but what about the green turtles who feed off the coast of Brazil but insist on laying their eggs 2000 kilometres away on Ascension Island?

We are always impressed with the ability of a species to buck the system with elaborate techniques of mimicry and camouflage or just straightforward cheating as in the case of the cuckoos. These cheeky creatures lay their eggs in another bird's nest and worse still when the young cuckoo hatches it proceeds to murder its foster brothers and sisters by pushing them out in the cold.

Ingenious examples of camouflage are plentiful. Some caterpillars take on the appearance of the type of vegetation which they live off, while the arctic fox which is white in winter but brown in summer is a relatively poor performer when compared to the chameleon which can change its colour to suit almost any background. There are butterflies who terrify their potential predators by doing a perfect imitation of an owl and others who take on the appearance of other poisonous butterflies even though they themselves would make a perfectly tasty meal. Practically all fish have developed a reflective surface which makes them indistinguishable from the shimmering light entering the water from above.

Altruistic behaviour among animals has always been of great interest to biologists and sociologists. Some creatures appear to put themselves at considerable risk in order to help their fellow creatures and among the social insects the vast majority of them appear to sacrifice themselves for the sake of others. The most common form of altruism is seen when a mother protects her offspring. I was privileged to observe a scene in a Malawian game reserve in which a female antelope feigned injury in order to lure a stalking leopard away from her offspring and towards herself. The trick worked and both mother and offspring escaped much to the leopard's chagrin.

Another of my favourite wild-life programmes concerned the miraculous behaviour of the lung fish. This ugly looking creature can survive the drying out of the river bed where it lives in Africa by burrowing into the mud and lying dormant until the water returns. The desiccated, emaciated fish has evolved the skill to breath oxygen and it is said that they can survive for years in prolonged periods of drought. When the rain comes, the fish instantly reverts to its normal appearance and habitat.

The reader may well wonder where all this is leading us. After all this book is about evolution and is not meant to be a study of wildlife. I have deliberately dwelt on the fascination of nature in all its varied forms because I want to emphasize and re-emphasize the point that nature appears to know no bounds when it comes to inventing survival strategies. The important thing to remember when looking at nature is that all these mechanisms of life are no more than a series of chemical reactions; what we observe in nature is no more than a demonstration of chemical possibility. It could almost be said that any useful survival mechanism which is chemically and physically possible has a chance of occurring in the course of evolution.

If, for a moment, we imagine a world without Darwin's Theory and set ourselves the task of speculating on the mechanism which produces the incredible variety of life we might well come to the conclusion that a mechanism other than random mutations and natural selection is responsible for this variety. Alternatively, having come to the conclusion, as I did, that Darwin's Theory is wrong on the grounds of improbability and absence of a convincing driving force, then we have no option but to speculate on a mechanism which is more likely than Darwin's Theory. The conclusion which I came to that day on the train between Piccadilly Circus and Charing Cross was this: if we can think of a mechanism for evolution which is superior to Darwin's Theory and which is chemically possible then we could have some confidence in predicting that such a mechanism already exists. Our confidence in the existence of the superior mechanism comes from our observations of nature which clearly demonstrates an almost limitless variety of chemical possibilities.

What would be this superior mechanism? In chapter two I argued that random mutations could not possibly produce the incredible variety which, according to Darwinism, acts as the raw material of evolution. We have no doubt that this variety does actually exist and therefore we must ask ourselves the following question; if the system for producing the variety is *not* random then what is it? The simple and obvious answer is that the system must be non-random. I believe that early on in evolution the inefficient mechanism described by Darwin's Theory would have been replaced by a mechanism which gave the primitive DNA an in-built ability to experiment with

45

different types of survival mechanism. Rather than wait for random mutations to supply the variety which is essential for survival the genome itself became a mechanism for change. As survival is dependant on flexibility then any mechanism which improves this flexibility will be favoured by evolution. If the ability to cope in the environment is the key to evolution then the slow and tedious method of waiting for random mutations would not have been the last word; a more efficient mechanism would have been found by nature to allow more rapid development in any useful direction. The new mechanism is certainly well within the bounds of chemical possibility when seen in the light of the mechanisms which we already know to exist and which already take us to the far reaches of our imaginations. Having rejected Darwinism the task before us is simple. We are faced with the facts of evolution and we must come up with a rational mechanism to explain those facts. What is the simplest mechanism which is chemically possible, which could explain the phenomenon of life? What are the probable ground rules of evolution?

A new theory would have to explain everything that we know to exist in nature. It would have to explain evolutionary progress and the existence of the incredible variety of life on earth. These are the two undisputed results of evolution. The rules of evolution would be very simple. They would be the minimum number of rules needed to do the job. We know from our knowledge of the simplicity of the DNA code that the mechanics of evolution do not waste time on unnecessary complexity. We have rejected random mutations as being the mechanism capable of producing the variety which we know to exist and therefore a cornerstone of any new theory must state that change is automatic and not random.

To produce variety on such a scale the genome would have to contain a mechanism which automatically creates future organisms which tend vary slightly from previous generations. The rules governing the amount and direction of this automatic change would have to account for the evolutionary progress which is apparent when life is viewed over millions of years. The new theory must also eliminate the strong link to the environment which is required by Darwinism. Change must happen without the scrupulous and constant examination by the environment. The new mechanism must be capable of producing altered working models (phenotypes) even before the phenotype has seen the light of day. Yes, the evolution of new bodies must take place without immediate reference to the outside world. Is this possible? Is this within the realms of chemical possibility? Yes, of course it is. Our everyday knowledge of evolution tells us that all organisms know much about the outside world in which they are about to exist without having seen it before. They can even be programmed to behave in such a way as to give the impression of having seen a video-recording of the outside world before they are born. By means of a pure chemical process an organism's behaviour is pre-programmed so as to give all the appearance of pre-knowledge. This knowledge of the world is stored within the chemistry of the genome.

The new theory, which we shall name the theory of the self-developing genome, states that the genome contains a program for designing new bodies which work perfectly in a world which they have never seen. The first rule of the new theory states that evolution proceeds automatically and the second rule states that the new working models, or phenotypes, are designed by the genome without immediate reference to the outside world. These simple rules are the basic minimum requirements that have made evolution possible and are a reasonable explanation of life on earth.

The most crucial change to our current understanding of evolution which I now ask the reader to accept is that life as we know it is a secondary-issue in evolution and the real phenomenon behind life on earth is the evolution of the mechanism which produces that life. Evolution discovered that the way to produce efficient organisms was to produce an efficient mechanism to experiment with different types of organism. Darwinism states that evolution acts on genes or individual organisms but the new theory postulates that evolution has acted on the mechanism for producing genes and organisms. By accepting that evolution acts at the production or design stage, then the whole task of creating variety in the finished product is accelerated to a phenomenal degree.

On a purely common-sense basis, what is required in evolutionary theory is a less than random mechanism which improves the odds in favour of useful or non-deleterious mutations. The current Darwinian view of evolution describes a mechanism in which the odds in favour of such mutations are zero.

The environment continues to play an important part in the new theory. It still plays a minor role in eliminating extremely unsuitable organisms but the environment has mainly acted on the rules or mechanism of evolution. The mechanism has been shaped by evolution to take account of broad environmental factors. The mechanism does its best to produce bodies which are capable of surviving. If the mechanism went out of control and produced hopeless bodies then natural selection would quickly eliminate that mechanism. The self-developing genome is little more than a set of mathematical rules for change. This automatic evolution could not be too abrupt otherwise the newly designed organism might not survive in its environment. Natural selection acts on the mechanism to encourage change within a certain range. In what direction is this change? If for the moment we can accept the concept of direction in evolution then the self-developing genome will produce change in any direction. For example, take a simple characteristic such as body size. The genome will automatically produce future generations that are either larger, smaller or the same size as the average of the present generation. Our old friend natural selection could encourage change in one or another direction but in the absence of any pressure from the environment the genome will automatically create variety in all and sundry directions. Another reason why the self-developing genome will not experiment with extreme degrees of change is that the new organism might be immediately eliminated in one generation because it is incompatible for sexual reproduction. What the new theory says is that offspring are *not* built from a faithful copy of parental genes but contain new genes automatically manufactured by the parental genes.

We know that the genome re-arranges itself during the process of sexual reproduction and the new theory merely states that this re-arrangement is subject to mathematical rules which ensures that the characteristics of future generations are able to evolve away from the characteristics of their ancestors. The mathematical rules, or algorithm, will not be simple, and successful change will not be easy to achieve. The successful re-design of a phenotype will still be a slow, tedious process but at least under the rules of the new theory new designs become a possible event rather than an impossible event. Reshuffling of the genome to produce useful evolutionary change might occur only after several consecutive re-arrangements as in a game of chess where a good player has to plan ahead with a variety of alternatives. The wonderful chemical program which builds bodies cannot easily be altered without mortally damaging itself.

To avoid any misunderstanding about the new theory we can draw parallels with modern day computer programming. There have been many attempts at simulating evolution via computer programs. Many biologists actually think that computer programs have already been written which demonstrate that it is quite possible to achieve change in a computer generated organism via a mechanism of random shuffling. But how exactly do these programs work? Essentially, programs have been written to generate a random output which can then act as the raw material of evolution. With all the other rules of evolution in place - death, sexual reproduction, species creation and natural selection, the artificial organisms generated inside the computer do indeed evolve. What then is the essential difference between these artificial programs for evolution and the genetic code? The difference lies in the fact that in the case of the computer simulation the program remains sacrosanct and unaltered; the program itself has been written to create variety but the program itself is not being re-arranged to create that variety. On the other hand the genetic program *itself* is being re-arranged. In the artificial world a program is creating something which is altered randomly but in the real world, according to Darwinism, we are expected to believe that the program itself is being altered randomly. Try for yourself to juggle the code or instructions of a computer program and see what happens!

You cannot juggle programs at random without harming them and therefore the genetic program, which can only create the raw material of evolution by re-arranging itself, must be re-arranging itself in an organised or algorithmic fashion. My message to all those "Eureka" shouting biologists who reckon that they have proved that computer programs have successfully simulated evolution is to say that all they have done is find evidence for the new theory of the self-developing genome; they have proved that the genome needs an algorithm to produce variety.

A glance back to chapter one and the six stages of evolution shows us that the unconscious aim of improving the ability of the organism to survive is achieved by increasing the flexibility of the organism. The improving scale of flexibility took us from the reliance on random mutations through the ability to instantly react to a stimulus to the accumulation and social transmission of survival tricks. The mechanism of the new theory is merely another survival strategy whereby the genome continually creates potential evolution which can be called upon when needed. If we accept our new definition of evolution as a progression through the six stages of chapter one then the crucial new point to take on board is that flexibility is the key to survival and not as Darwinians would have us believe a perfect fit in a particular environment. The last thing that organisms need is to be specialists in a narrow environment; specialization is the road to extinction in evolution, flexibility is the real key to survival. The self-developing genome gave life on earth the flexibility it needs to survive.

Our concept of the genome will have to change to accommodate the new theory. We will have to dispense with the idea that the millions of genes which are strung along the DNA spiral staircase are simply a one-for-one means of creating other chemicals which build the immediate body. When our understanding of the genome is improved we will discover a multi-hierarchical, non-linear computer program which is only partially a builder of bodies but is also a plan of immense complexity and ingenuity which experiments with form within the limitations of a framework based on the whole range of stimuli and conditions which exist on earth. New species will be formed with the mating of two individuals who have diverged sufficiently to make them reproductively isolated from the rest of the population. If the direction of evolution proves disadvantageous then natural selection will play the part of executioner and the direction could be altered but it is just as likely that a newly developing species will search for a new niche in which it feels more comfortable.

I did say in the last chapter that we would come back to the phenomenon of "pre-adaptation" and it can be seen that under the new theory pre-adaptation is more the norm than the exception to the rule. Life is continually experimenting with form without the stimulus or pressure from the environment and if vacant niches are available then the new forms of life will seek out those niches to which they are best suited.

I find the view of my garden from my study so much easier to understand in the light of the new theory. Does that tree or that bush or those blades of grass really need to be like that to survive? Is there really any significance to the random variety of shapes which we see amongst leaves? The answer is no. They have taken on that form because it is one of many forms available to life on earth; forms which are not so unsuitable as to give the organism a disadvantage but on the other hand are not so advantageous that a significant link with environment can be proved.

Another consequence of the new theory is that the availability of niches will play a part in encouraging diversification. In a crowded environmental situation natural selection might have the effect of discouraging the manifestation of new forms, whereas an "explosion" of new forms could occur in the opposite situation. In fact research by Dr. Wayne Sousa of the University of California shows that there is a greater variety of species on rocks which are occasionally subject to stormy seas than on rocks more sheltered from the elements. The storm swilled rocks have more vacant niches than the less disturbed sheltered rocks.

Not only the view from my garden but the whole of nature is well explained by the new theory. Life went from swimming in water, to crawling on land to flying in the air because of the inbuilt propensity of the genome to experiment with forms which could take advantage of the physical conditions which exist on earth.

The incredible radiation of evolution which results in an endless variety of creatures surely cannot be explained in terms of individual responses to environmental pressures. It is easier to understand evolution as a random experimentation with form, filling niches whenever available and changing direction without regard to the environment. The ever continuing experimentation can lead down unexpected pathways. Birds which could once fly become earth bound, mammals which were once terrestrial return to the sea and other creatures seek out the most hazardous environments in order to fill an empty niche.

The radiation effect of evolution makes me think of the evidence which led cosmologists to the Big Bang Theory. All the galaxies in the universe appear to be moving apart as if they are on the surface of an inflating balloon. Evolution is proceeding in a similar manner although the tendency for everything to evolve away from everything else is subject to the restrictions imposed by the environment, which unlike the expanding universe, becomes crowded and competitive.

The pattern achieved by the radiation is a mathematical consequence of the continual experimentation and similarly the rate of radiation is also subject to mathematical rules. In his book *The Great Evolution Mystery*, Gordon Rattray Taylor says "we find some lines which evolve very slowly if at all, and others which evolve very rapidly, with a much larger group between the extremes. In this middle group there is a certain spread of rates, within each phylum, but the very slow and the very fast are outside this range." This description of the creation of new species fits very well with the idea of the self-developing genome.

Natural selection still plays a part in evolution but no longer as the main mechanism. In the early primordial days of primitive organisms it was pure unadulterated natural selection which decided the fate of the species. Evolution in those far off days was painstakingly laborious, relying on the incredibly slow accumulation of random mutations to induce change. What was needed by evolution was a mutation which allowed the genome to develop under its own steam, a mutation which was initially created by natural selection but which usurped natural selection as the main driving force of evolution. In the final analysis the new mechanism of the self-developing genome is still under the control of natural selection. Experimentation with forms which are unsuitable for the conditions which the

53

organism can find on earth will be discouraged or eliminated by natural selection. Experimentation would be subject to rules and limitations; it is no use, for example, experimenting with conditions which are only found on Jupiter. The self-developing genome was originally evolved by natural selection and it will continue to be guided by and improved upon by natural selection. Natural selection now takes on a new significance; it is responsible for the evolution of the mechanism for creating variety. The new mechanism takes over from the Darwinian view of natural selection and acts to accelerate the production of viable alternatives.

The element of chance is largely removed by the new theory. The various working parts of the forms which the self-developing genome is able to express are all well coordinated with each other; the preposterous idea of one favourable mutation waiting for another favourable mutation in order to achieve compatibility between all the organs is eliminated with the new theory; the long neck of the giraffe would have killed the animal if it hadn't simultaneously developed a powerful pump to get blood to its brain.

Gordon Rattray Taylor also thinks that Darwinism is inadequate to account for the inter-relationship of all the working parts of a phenotype. When discussing the number of coordinated changes needed for an adaptation such as colour change or blood production he comments, "That these sequences of coordinated reactions - and there are literally thousands of them in the human body - should all have arisen by random mutations of single genes is in the highest degree unlikely." He goes on to quote T.H.Frazzetta of the University of Illinois as saying, "There is a definite suggestion that certain modifications cannot greatly precede or lag behind others but must keep pace if the performance of the machine is not to become sloppy."

I do not think that there is any room for coordinated changes in Darwin's Theory and our limited understanding of the genome does not help to throw any light on this subject. A controversy in the weekly periodical Nature however pointed to some interesting research which seems to suggest that coordination or anticipation in evolution has been detected. Commenting on a paper by Barry Hall, Neville Symonds writes, "in Hall's system, two independent mutations are required before bacteria can adapt to a novel environment. Neither mutation on its own is expected to confer any adaptive advantage, yet in the presence of selection the double mutants occur at frequencies orders of magnitude higher than expected from the individual spontaneous rates."

We must never forget the tremendous information storage potential of the genome. We understand very little of the genetic program at present but it is inevitable that much of the chemical program is given towards coordinating the building of bodies. We will no doubt eventually discover the kind of repetitive "loops" which are common in our own computer programs, where a small section of the program can be repeatedly used. For example, it would be surprising if those parts of the program for building multiple vertebrae, feathers, scales, teeth, fingers etc. were entirely unconnected. We might eventually understand the uncanny symmetry of living things; it is inconceivable that evolution had to invent two eyes via two separate routes for those organisms which possess binocular vision. The genome will eventually surprise even its greatest admirers when the full extent of its ingenuity is revealed.

It strikes me that Darwin's Theory has a built-in self-destruct button hidden within its own definition. My argument for this assertion runs something like this: if the genome really *is* the inexhaustible source of the raw material which has been needed by evolution then we should look to that source as the main agent of evolutionary change and not to natural selection. The process which Darwinists agree is the original source of variety must be such an incredibly reliable and efficient mechanism that the process itself should be seen as the main driving force of evolution and not the mechanism of natural selection which is an event which follows the main act of creating perfect working organisms.

If the unconscious aim of evolution is an improved survival strategy then the genome has an unconscious knowledge of its organism's body and the conditions in which the organism will survive. The feedback to the genome is via natural selection; in the same way that experimenting with extreme forms incompatible with the conditions on earth will be eliminated by natural selection, so will forms where the various working parts are badly coordinated.

The strict dividing lines between species is unaffected by the new theory; sexual reproduction is the mechanism which isolates one species from another. The bewildering array of species which are known to currently exist on earth which numbered several million at the last count are but a minute proportion of the total number which have ever existed. It is a lot easier to comprehend this immense variety of form when seen in the light of the new theory. The great number of known species no longer needs to be explained in terms of each species trying to adapt to its particular environment but it is merely the mathematical consequence of a mechanism which manufactures variety. The Kestrel developed its flying skill and keen eyesight before it found its niche for hunting. The reason we observe numerous different species occupying the same territory is not due to a response to environmental pressure but is a result of millions of years of experimentation with form. The bewildering assortment of colour and shapes which cohabit on a coral reef is a fine example of the random variety of evolution.

The study of evolutionary genetics is largely unaffected by the new theory. The new theory, after all, merely changes the probabilities of mutations happening; the same sort of mutations will happen whether the evolutionary game is played under Darwinian rules or under the new rules of the self-developing genome. Evolutionary genetics is largely a theoretical subject which is concerned with explaining evolutionary changes in terms of mathematical models. Simplistic models are constructed to predict the likely frequency of a particular gene over a specified number of generations in the gene pool of a species. The reason why mathematics has been such an ideal tool with which to investigate evolution is that biologists are fairly convinced that all the parameters on which the mathematical models are based are known. The most important parameter concerns the fact that sexually reproducing organisms contain one set of genes inherited from the female parent and one set from the male. Each gene from the male exists in juxtaposition with an equivalent gene from the female but the crucial point is that only one of these alternative genes will be used by the organism to build its phenotype. Sometimes the alternative genes are identical, in which case it doesn't matter which one is used but often the two genes are different and it is the *dominant* gene which is expressed. When the gene pool as a whole is considered then for any particular position in the genome there may be many different possible genes in either the male or the female set, and thus the gene pool is a store of variety which can be explored with mathematical models. The variety manufactured by the self-developing genome need not be immediately expressed in a finished body but could be hidden by the phenomenon of dominant and recessive genes.

If we return to the question of mutations, a theory already exists which argues that random mutations are not very often subject to immediate scrutiny by the environment via natural selection but are said to be "neutral" with regard to the environment. This build up of neutral mutations in the gene pool can lead to a random drift in the direction of evolution. The new theory of the self-developing genome embraces the theory of neutral mutations wholeheartedly but the two theories have little in common except their assertion that changes within the genome can occur without being constantly vetted by the environment. In the theory of the self-developing genome evolution has given the genome the skill to experiment, within a certain range, with all the physical characteristics of the organism. The changes which are being planned by the self-developing genome are well coordinated moves about the norm whereas the theory of neutral mutations still relies on the Darwinian assumption of completely random mutations of genes being the source of variation.

There is no definite proof available for the new theory in the same way that there is no definite proof of Darwin's Theory. Evolutionary theories are argued out using whatever evidence is available in a similar way to a legal case being argued out in court between contesting lawyers. Given a lack of definite proof you must agree that the new theory so far fits the evidence better than Darwin's Theory. The correct theory of evolution will only be finally revealed when we can completely unravel and understand the chemistry of life. It might eventually need several textbooks listing nothing but chemical reactions to understand all the stages in the synthesis of a human being but I believe we *will* some day understand every step in the process and thereby demonstrate one way or another how evolution proceeds.

The essential point about the self-developing genome is that offspring will tend to vary from their parents in a positive or negative evolutionary direction. There would be a limit to the degree of variation and the offspring could vary by a degree which is anywhere between the negative and positive limits including zero variation. The use of the terms positive and negative are not meant to signify forwards or backwards or better or worse but are merely meant to signify different directions. In contrast to the self-developing genome theory, Darwin's Theory would predict, in the absence of random mutations, no evolutionary change from one generation to the next.

Finding evidence for the new theory will lend it credence, and the next chapter is devoted to a search for that evidence. We will also see if the new theory fares any better than Darwin's Theory when trying to explain some the outstanding mysteries of evolution.

Chapter Four

A New Look at Some of the Unresolved Problems of Evolution

The trillions of different individual organisms which we know to have existed on earth since life first appeared more than 3.5 billion years ago are not the result of the sifting effect of natural selection acting on variety created by random mutations. The creation of variety now and in the past goes on as relentlessly and as reliably as the rising and the setting of the sun. The production of this variety is automatic and unstoppable. From the known facts of evolution we have to conclude that the raison-d'etre of the genome is to create variety. The creation of such variety is good for the survival chances of the organism and has the inevitable side-effect of evolutionary progress that eventually led to human beings.

When reading some of the voluminous literature on modern research undertaken in the fields of genetics and microbiology I cannot help but feel that Darwin has already been disposed of without anybody noticing. Substantial amounts of new knowledge sound very unDarwinian to me. Before we have a look at some of the findings of this research it may be a good point to summarize how the DNA chain starts a process that leads to the manufacture of bodies.

It was briefly mentioned in chapter two that genes consist of a ladder-like structure whose rungs are composed of four different chemical bases, adenine, guanine, cytosine and thymine. In fact each rung consists of a pair of bases but the number of different base pairs remains at four because adenine always pairs with thymine and cytosine always pairs with guanine. It is this property of the double stranded DNA which allows it to faithfully reproduce itself after splitting down the middle to form two single strands. The other important feature of DNA is that it is not single bases which constitute a letter in the coding alphabet but a specific group of three bases, a triplet. The number of letters in the alphabet is therefore 64 (i.e. 4 times 4 times 4) and many of these 64 letters comprise a code which instigate a series of complex chemical reactions which in turn lead to the synthesis of another chemical known as an amino acid. The amino acids are the building blocks of proteins and it is the three dimensional spatial structure of proteins which give all the cells in our body their particular characteristics and physical properties. There are in fact only 20 different amino acids and therefore some of the 64 letters of the DNA alphabet are redundant. The twenty different amino acids can be permutated into an almost unlimited variety of different chains of proteins of various length. The average protein is made up of between 100 and 500 amino acids and an individual gene is often defined as a length of DNA which can code for one particular protein. The coding system is extremely economical in that four different bases and an alphabet made of triplets is the simplest system which could satisfy the twin requirements of reproduction and protein synthesis.

A typical genome consists of between a few million and a few billion base pairs but geneticists estimate that as few as 5% of the triplets play a part in protein synthesis. If the researchers are looking for a use for the other 95% then might I suggest that some cognizance is taken of the new theory proposed in this book.

From what we know of DNA it appears to use the minimum of chemical and arithmetical complexity to achieve its twin tasks of building bodies and making copies of itself. There is little waste. I would guess that there is also little waste in the amount of DNA stored in the genome. Even though the function of the vast majority of the DNA is unknown to us I suspect that eventually we will see that nearly all the DNA plays a part in evolution. An obvious consequence of the new theory is that some of the genetic program is concerned not only with building the immediate body of the organism but it is also concerned with planning for future change.

Despite the 95% of the genome which is not understood by geneticists, modern research techniques are throwing light on a genome which is revealing itself to be rather ingenious. The focus of attention for geneticists are individual genes and we ought to pause for a minute and consider what is actually meant by the term "gene." An individual gene is loosely regarded as an indivisible unit of inheritance and is composed of hundreds or thousands of base pairs. The gene therefore is a collection of base pairs which is passed from generation to generation as a coherent unit. In the same way that the DNA coding structure is extremely economical we are finding that genes are also extremely efficient in their operation. We have known for a long time that physical characteristics in the organism are often controlled by several genes working in harmony but more recent discoveries have thrown light on the ability of a single gene to influence more than one characteristic. A further example of the economy of genes came to light when it was discovered that certain strings of bases could be read in more than one way depending on the point where you started counting the triplets. In other words a sequence of the four different bases such as ATCTCGTAGCTTCAC would normally constitute the series of triplets ATC-TCG-TAG-CTT-CAC which is the code for a particular group of amino acids. If however the first "A" base was ignored the remaining sequence of bases would constitute a different series of triplets TCT-CGT-AGC-TTC which would code for a different group of amino acids. Each different interpretation of the sequence was shown to play an important part in protein synthesis and the achievement of such ingenious economy within the genome would astound our most advanced computer programmers.

Further clues to the complexity of operations within the genome have come with the discovery that not all genes are equal when it comes to the part they play in the protein manufacturing process. Some genes rely on other genes to tell them when to start and stop manufacture and more senior genes appear to control the behaviour of junior genes in a strictly hierarchical way. Geneticists have succeeded in growing legs in the wrong part of a fly's body by fooling the gene which has overall control of leg production to start work. The other subservient genes in leg production don't mind where they are on the fly's body, they just obey the orders of the boss gene. The boss genes are sometimes responsible for controlling the switching on and off of whole batteries of genes in order to achieve the coordinated construction of the organism.

The hierarchical structure within the world of genes was further demonstrated when is was discovered that certain genes seem to play a part in encouraging the rearrangement of other genes during sexual reproduction. Under the rules of both the new theory and Darwin's Theory this constant rearrangement of genes within the genome creates the variety which evolution needs. Other clues pointing to the ability of the genome to reorganise itself have been known for some time. There are sections of DNA which are specialists in moving up and down the spiral structure by breaking the chain, inserting themselves and finally repairing the damage by relinking the separated ends. Other methods of self reorganization include the duplication, elimination or inversion of whole sections of DNA.

The point which I am trying to make by this extremely superficial and brief look at some of the results of genetic research is that the genome is proving to be well equipped to experiment with the vast array of possibilities open to it. The experimentation is carried out without reference to the outside world and the only feedback received by the genome would be when the organism developed a characteristic which was highly unsuitable for any available niche. The new theory states that the mathematical rules which control the rearrangement of the genetic code have evolved via this feedback. There is nothing revealed by modern research which could definitely discount the possibility that the new theory of the self-developing genome is correct. In fact the evidence points to a genome of immense skill rather than a genome whose main function is the preservation and duplication of its genes.

The difficulties presented to researchers in their endeavour to understand the genome have encouraged many microbiologists to concentrate their energies on studying the proteins which are constructed from the DNA program. You will recall that proteins are constructed from a chain of amino acids and it was thought that a study of this sequence of amino acids might shed some light on the evolutionary history of a particular protein.

It was discovered in the late 1950s that the same protein which exists in a number of different animals is in fact measurably different when it is analysed into its sequence of amino acids. For example, the protein cytochrome C which occurs in a wide variety of life-forms from bacteria to humans is revealed to have a great deal of inter-species variety when amino acid sequences are compared. When comparing the cytochrome C in different species the physical properties and characteristics of the protein appear to be indistinguishable but when viewed at the molecular level there is sufficient variation to establish an accurate measure of difference or relatedness between the species. This quantification is a measure of the number of positions in the amino acid chain which differ when comparing the protein in one species with the same protein in another species. It was soon discovered that these comparisons when made for large numbers of species might be a guide to evolutionary relationships. The science of comparing different species with the aim of classifying organisms into various groups is called taxonomy and this attempted classification was historically based on a comparison of the physical characteristics of each species and is often open to vigourous disagreement. With the new science of protein sequencing biologists at last had a method of making comparisons using a mathematically based technique which was not subject to human error or bias. Fortunately for taxonomists from the ancient Greeks to the present day the relationships between species as revealed by protein sequencing are almost identical to the relationships as determined in the more traditional way. Minor differences between the two approaches have sometimes led to a reassessment of a view held by the traditionalists but by and large the degree of consensus is remarkable.

The other biological discipline for which protein sequencing has an important bearing is Palaeontology or the study of fossils. An understanding of the geological age of rocks enabled Palaeontologists to construct a "tree of life" showing the order in which the various groups and species appeared and to draw conclusions as to what evolved from what and when.

The matrix on page 66 has been reproduced from Michael Denton's book *Evolution: A Theory in Crisis*, and it is an extract from The Dayhoff Atlas of Protein Structure and Function. Each figure in the matrix represents the percentage by which the amino acid sequence of cytochrome C differs between any two species. For example, the cytochrome C of Carp differs by 25% from the cytochrome C of the Silkworm. The matrix gives support to the palaeontologist's view that bacteria came first, followed by plants, insects, fish, birds and mammals. There is however something very significant about the statistics in the matrix. A closer look at the relatedness of the bacteria Rhodospirillumi rubrum C2 shows that it is as different from a horn worm as it is from a horse yet the palaeontologist's tree of life shows that the horn worm was evolved at an intermediate stage between bacteria and mammals. Why doesn't the technique of protein

BACTERIA · YEASTS · PLANTS · INSECTS · MAMMALS, BIRDS, REPTILES, TELEOSTS, CYCLOSTOMES

Column order (left → right): Rhodospirillum rubrum C₂ | Baker's Yeast | Debaryomyces kloeckeri | Candida krusei | Wheat | Sunflower | Castor | Tobacco Horn Worm Moth | Silkworm Moth | Screw-worm Fly | Lamprey | Carp | Bonito | Tuna Fish | Snapping Turtle | Pigeon | Pekin Duck | Penguin | Kangaroo | Dog | Horse

	R. rubrum C₂	Baker's Yeast	D. kloeckeri	C. krusei	Wheat	Sunflower	Castor	Horn Worm	Silkworm	Screw-worm	Lamprey	Carp	Bonito	Tuna Fish	Turtle	Pigeon	Pekin Duck	Penguin	Kangaroo	Dog	Horse
Horse	64	42	42	46	41	41	40	26	27	20	15	13	17	18	11	11	10	12	7	6	0
Dog	65	41	41	45	39	39	38	23	23	19	13	11	16	17	9	9	8	10	7	0	6
Kangaroo	66	41	42	46	42	39	38	26	26	22	16	13	17	17	11	11	10	10	0	7	7
Penguin	64	42	42	45	41	40	38	25	25	22	18	14	17	17	8	4	3	0	10	10	12
Pekin Duck	64	41	42	45	41	39	38	25	25	21	18	14	17	16	7	3	0	3	10	8	10
Pigeon	64	43	42	45	41	39	38	24	25	20	18	13	17	17	8	0	3	4	11	9	11
Turtle	64	45	41	47	41	39	38	27	26	22	18	13	16	17	0	8	7	8	11	9	11
Tuna	65	44	43	42	43	44	45	28	30	22	18	8	2	0	17	17	16	17	17	17	18
Bonito	64	42	42	42	41	41	41	29	31	23	18	7	0	2	16	17	16	17	17	16	17
Carp	64	44	41	45	42	41	41	24	25	20	12	0	7	8	13	13	14	14	13	11	13
Lamprey	66	45	45	50	46	44	45	30	31	26	0	12	18	18	18	18	18	18	16	13	15
Screw-worm	64	40	39	43	40	40	40	11	13	0	26	20	23	22	22	20	21	22	22	19	20
Silkworm	65	42	39	43	40	40	40	5	0	13	31	25	31	30	26	25	25	25	26	23	27
Horn Worm	64	44	39	42	38	39	39	0	5	11	30	24	29	28	27	24	25	25	26	23	26
Castor	66	43	43	45	12	10	0	39	40	40	45	41	41	45	38	38	38	38	38	38	40
Sunflower	67	43	43	44	13	0	10	39	40	40	44	41	41	44	39	39	39	40	39	39	41
Wheat	66	41	41	42	0	13	12	38	40	40	46	42	41	43	41	41	41	41	42	39	41
C. krusei	72	25	23	0	42	44	45	42	43	43	50	45	42	42	47	45	45	45	46	45	46
D. kloeckeri	67	27	0	23	41	43	43	39	39	39	45	41	42	43	41	42	42	42	42	41	42
Yeast	69	0	27	25	41	43	43	44	42	40	45	44	42	44	45	43	41	42	41	41	42
R. rubrum C₂	0	69	67	72	66	67	66	64	65	64	66	64	64	65	64	64	64	64	66	65	64

The Cytochrome Percent Sequence Difference Matrix. (from Dayhoff)

sequencing reflect this intermediacy? In fact the protein sequencing indicates that the bacteria is almost equally removed from all other species; no species can be seen to be intermediate between the bacteria and any other species. A similar argument can be applied to other statistics in the matrix. For example, although it is evident from the statistics that castor appeared after yeast in the evolutionary tree of life and is even further removed from bacteria, it is impossible to rank the species which came after castor when comparing their protein sequence to that of castor. In other words all species evolving after castor appear to be equidistant from castor according to protein sequencing. What logical conclusion can be drawn from the statistical pattern which emerges from this matrix?

Before drawing our conclusions it is important to remind the reader of one of the important prerequisites outlined in chapter one which had to be accepted before setting out to search for the likely mechanism of evolution. In chapter one it was stated that we shall assume that evolution has in fact happened. This assumption is based on the evidence of the fossil record and our interpretation of the results of protein sequencing are based on the assumption that the fossil record is a reasonable guide to the evolution of life on earth. Michael Denton quite correctly points out that protein sequencing cannot be used as proof that evolution has happened but it can give us pointers to the way in which evolution occurs. Now let us return to the matrix.

The first conclusion to be drawn from the data contained within the matrix is that the evolution of proteins is an ongoing process which continues in ancient groups such as bacteria as well as modern groups such as mammals. The protein of ancient wheat has evolved into the protein of modern wheat even though the naturally occurring versions of old and new wheat do not demonstrate very much evolution of their bodies or phenotypes. Evolution at the molecular level is proceeding even though evolution of the phenotype appears to be stagnant. This is excellent support for the theory of the self-developing genome which predicts that evolutionary change can remain hidden within the gene-pool. Evolution is proceeding under its own steam without reference to the environment.

We might pause for a moment to reflect on a very important aspect of evolution. What exactly is evolving? From a human perspective most of us like to think of the organism as being the object of evolution. But it is just as valid to observe evolution acting on individual cells that make up the body or even individual genes which constitute the program that builds the body. In the other direction, family units and societies can be seen to evolve. Viewed at the level of individual proteins cytochrome C can be seen to be trying to evolve under the pressure of the self-developing genome. The environment or niche of cytochrome C is very narrow and rigid. The evolution of its physical structure is greatly restricted. Although its genetic code is changing relentlessly natural selection has severely restricted any alteration in its physical form or its chemical behaviour. Any change at the phenotypic level would be disastrous for both the body which houses the protein and the protein itself.

What else can be gleaned from the statistics reproduced on page 66? What about the rate of evolution? The conclusion here is even more astounding. If the protein cytochrome C of Rhodospirillumi rubrum C2 is as equally distant from the same protein in all species which evolved later then we must conclude that the rate of evolution of this protein is constant with regard to time. This unavoidable conclusion is truly amazing, especially when you consider that the organisms listed in the matrix have widely differing reproduction rates. Evolution appears be proceeding even during the lifetime of the organism! This is a very difficult concept to explain but the point to make is that evolution at the molecular level proceeds automatically and is not under the extraneous influence of the environment. This is further support for the new theory.

Darwinists manage to explain away the phenomenon of the "molecular clock" as follows: although they accept that non-deleterious re-arrangements of the genetic code occur on a regular basis this is not necessarily reflected in the evolution of the organism or phenotype itself which only evolves on a non-regular basis via natural selection. My answer to them is this: why is it that the taxonomic trees of life as drawn up by Aristotle onwards which are based on observations of the phenotype, are exactly the same as those drawn from the results of genetic sequencing? If we accept that the genetic sequences are evolving on a regular basis then we must assume that evolution as measured in the traditional way is proceeding at the same rate.

Where does this leave Darwin's Theory? Nowhere I'm delighted to say. The idea that evolution proceeds at a constant rate was anathema to Darwin; "It has been erroneously asserted that the element of time has been assumed by me to play an all-important part in modifying species, as if all the forms of life were necessarily undergoing change through some innate law." If evolution was driven by changes in the environment as Mr Darwin insisted then you would not expect the rate of evolution to be constant. The theory of natural selection continues to play its part in fine tuning the organism to its environment but it is no longer needed as the driving force of evolution because the molecular evidence demonstrates that evolution proceeds under its own steam. Real evolution is ticking away in the genome without any immediate reference to the environment and is storing up a variety of characteristics within the gene pool.

Although biologists have not managed to explain the molecular clock in terms of Darwinism some have at least recognised its existence and have used it to determine the point in time when two separate species evolved from a common ancestor. One of the more interesting conclusions from this method of dating the branching of the evolutionary tree is that humans and chimpanzees evolved from a common ancestor only five million years ago and not fifteen million years as was previously thought.

The fossil record re-examined

The evidence which led eighteenth and nineteenth century naturalists to doubt the traditional view that life on earth had been created instantly by God came from the fossil record which indicated that life had evolved slowly over millions of years. Our vast collections of fossils show how the various groups and individual species evolved over geological time and using our knowledge of the age of the rocks which preserved the fossils we are able to arrange all life forms in order of appearance. For example, the fossil record points to the fact that invertebrates were abundant 600 million years ago and the first fish appeared 500 million years ago. Reptiles evolved 150 million years after the first fish and mammals appeared on the scene 175 million years ago. A closer look at the fossil record reveals two aspects of evolution which have been the subject of great debate among biologists. The first argument concerns the mystery of the period 600-700 million years ago which gave rise to the sudden appearance of the full range of invertebrates. Remember the earliest fossil record of life on earth takes us back 3500 million years ago but we had to wait another 2800 million years before anything more interesting than algae or bacteria appeared. After the appearance of the invertebrates it was only another 400-500 million years before all the major groups which exist today were evolved. Why did evolution take so long to get going and then suddenly change gear to allow this comparatively rapid evolution of more sophisticated life forms? The geological period which started 600 million years ago is known as the Cambrian and the period of rapid evolution following the 2800 million years of slow evolution is known as the pre-Cambrian explosion.

Biologists have given several possible reasons for the paucity of fossils in rocks older than 700 million years. It has been suggested that these older rocks were not suitable for preserving fossilised remains of anything but bacteria and algae. Another school of thought says that conditions on earth might not have been conducive to the evolution of the more sophisticated life forms until the time of the pre-Cambrian explosion. These and other explanations have been largely discounted by the academic world and this slow start to evolution has remained a complete mystery. The new theory of the self-developing genome is now put forward as a possible explanation of this change of pace of evolution. I freely admit that I have no proof for this claim but even without the new theory it strikes me as being a perfectly reasonable assumption that the pre-Cambrian explosion was the result of a change in the mechanism of evolution and this new mechanism was more efficient at creating new life forms than the old mechanism. One of the obvious consequences of the self-developing genome is the more rapid pace of evolution. Evolution no longer waited for the unlikely occurrence of favourable mutations but could now be driven by a genome devoted to experimentation with perfectly coordinated finished products. I am not, quite frankly, convinced that the self-developing genome suddenly appeared on the scene 700 million years ago. The self-developing genome, like everything else in evolution, evolved slowly, but the pre-Cambrian explosion might have been the result of an improved version of the self-developing genome. On the other hand the change in pace of evolution might have been due to another facet of the evolutionary mechanism such as the proliferation of sexual reproduction as the most common means of propagation. Even if the self-developing genome was not responsible for the pre-Cambrian explosion I am still surprised that a change in the mechanism of evolution has not been suggested as an explanation. This is a consequence of a strict adherence to the Darwinian doctrine; most biologists accept that Darwin's Theory is essentially correct and therefore alternative theories are never discussed when trying to account for some of the unexplained mysteries of evolution.

Many clues to the mystery of the wonderful workings of nature can often be found in technical advances made by man. Machines and computers are constantly evolving in order to improve their efficiency. Computers will eventually work like a human brain; nothing could be more efficient at processing information. The race to understand the workings of the brain will be won by computer engineers and not by neurologists. Computer chips are now so complex in structure that they are often designed by computers. Yes, computers are now used to design computers. Eventually computers that design computers will be designed by computers. It is a fair bet that this progress in computer engineering is following a course long since discovered by nature. The self-developing genome is a genetic program which designs genetic programs.

Another problem with the fossil record is the absence of fossils showing the gradual evolution of one species into another. Instead, what we see is evidence of jerky evolution, new species appearing suddenly without the numerous intermediaries which we would expect to find according to Darwin's Theory. Darwin told us that organisms will evolve by the accumulation of innumerable slight modifications and he abhorred the idea of any sudden steps in evolution. Darwin assumed that the lack of intermediate forms was due to the "imperfection of the geological record" and he was hopeful that a more complete picture would emerge with the discovery of more fossils. The continuing fossil search has not in fact helped and the record today is as jumpy as it was in Darwin's day. In fact the vast quantities of fossils discovered since *The Origin of Species* was published has emphasized the clear steps in evolution (at least at the phenotypic or morphological level) rather than helped to smooth it out. Some biologists now accept that evolution is not as smooth as Darwin would have us believe and the search for an explanation of this "jerkiness" has led to a fierce debate among leading contemporary writers on evolutionary theory.

The most convincing explanation of the lack of intermediate forms is that rapid evolutionary change occurs when a small proportion of a population of a species becomes geographically isolated and is subsequently subject to different environmental pressures from the rest of the population. New useful genes would spread far quicker in a small population and the combined factors of speed and low population would lead to a lack of fossil remains. Notwithstanding this satisfactory explanation, the theory of the self-developing genome also explains the missing links in evolution. Fossilised remains are a record of phenotypes and not genomes and the theory of the self-developing genome tells us that evolutionary change can occur within the gene pool without necessarily affecting the phenotype. Because of the phenomenon of dominant and recessive genes evolutionary change within the gene pool can remain hidden from fossil hunters and the change which they perceive in the phenotype is not necessarily a reflection of the true rate of evolution of the genes. The gene pool can store up evolutionary change which can be rapidly expressed in the phenotype when needed. The greatest drawback to the theory of rapid evolution in small isolated pockets is that the mechanism still relies on random mutations whereas with the new theory rapid evolution is possible because the new forms are ready and raring to go.

Under the influence of the self-developing genome the degree by which one generation differs from another, phenotypically, is restricted for two reasons. Firstly change has to be within a narrow band otherwise offspring could be produced which are incompatible with any available niche and secondly any offspring which was too different from other members of its species might be incompatible for sexual reproduction. In these cases the genes responsible for this speedy evolution would tend to be eliminated. But occasionally the rule will be broken. A significantly different offspring will sometimes survive which, although it might be incompatible with 99.9% of other members of its species, might occasionally be able to successfully mate and the offspring resulting from that liaison might then be the first generation of a new species. The self-developing genome is constantly experimenting with form but most new features will be averaged back into the population via sexual reproduction. Occasionally though, a new, significantly different freak will survive. It is all a matter of probabilities.

A new species is obviously created from just two individuals and therefore their offspring will be very rare until the population increases. When the population is large enough to constitute an established species then the self-developing genome would take it, in evolutionary terms, even further away from its parental species. The laying down of a fossil is a rare event and it is not surprising that whilst a new species is building up its population base then very few fossils are laid down for future discovery.

Another way of looking at the phenomenon of the self-developing genome is to imagine a genome desperately trying to evolve but being held back by natural selection at the phenotypic level. Experimentation is being curtailed by either a lack of compatible mates or available niches but occasionally pressure from the self-developing genome bursts through these restrictions and a new species is born. With luck the new species will manage to hold onto existence in extremely small numbers but eventually might become sufficiently established to leave fossil remains.

Although the above application of the new theory points

to evolution of the phenotype proceeding at a rather uneven pace, the long term effect of the self-developing genome on the store of form within the gene pool will lead to a strong correlation between the evolution of the gene pool and the evolution of the phenotype. The phenomenon of species creation coupled with the random search of the self-developing genome will inevitably produce sufficient divergence to give the impression of progress and increasing complexity. This divergence could also give rise to stagnation and the occasional step backwards but natural selection will ensure that the route taken by this random drift is not disastrously disadvantageous to the species.

The randomness of evolutionary progress which is a consequence of the new theory can be seen clearly from a taxonomic view-point. Taxonomists have a classification system which starts at the level of individual species and then moves up one level to a genus which represents a group of related species and then on to a family, an order, followed by a class, a phylum and finally the three kingdoms of plants, animals and fungi. For example, the tiger is a species in its own right which belongs to the genus Panthera, which in turn belongs to the family Felidae which is part of the order Carnivora which is a member of the class Mammalia, which is included in the phylum Vertebrata which belongs to the Animal kingdom. The pattern which emerges from taxonomy shows for example that species radiate from genera (plural of genus) on a random basis; some genera having many species and others having only one. The rate and diversity of evolution can be shown to be regular with regard to time and regular with regard to branching; there is no indication that the rate of evolution or its diversity is under the control of an unpredictable pressure from an external source such as the environment.

A further great unsolved mystery of evolution concerns the phenomenon of parallel evolution. This is when two distinct species who have not had a common ancestor for millions of years and who are geographically isolated from each other evolve similar characteristics. Organs such as the eye which are common to many species can be shown to have evolved independently on a number of different occasions. This example of parallel evolution is perhaps understandable given the usefulness of light perception and the lack of viable alternatives open to evolution when trying to develop a light sensitive system. Similarly a particular beak or claw could evolve more than once in response to the same need. The new theory copes very well with this type of parallel evolution because the genome has an unconscious knowledge of the outside world; it will experiment with bodily structures that can cope with the sort of stimuli which the organism will face on earth. The self-developing genome has the effect of speeding up evolution compared to the inconceivably slow route of waiting for useful random changes.

Other types of parallel evolution are not so easy to explain in terms of a similar response to a similar need. For example, the marsupial wolf evolved independently from the mammal wolf and their last common ancestor is thought to be a creature which had no wolf-like characteristics at all. The two wolves would not have been given the same name if it was not for their uncanny physical similarities but it is difficult to understand why the two creatures evolved such similar characteristics, especially when you consider the vast differences in their immediate environments. I have not drawn any specific conclusion from the many examples of this type of parallel evolution which we know to have occurred. The reader might think that the common ancestor of the marsupial and mammal wolf might have had wolf-like features hidden within its gene pool and therefore the phenomenon of parallel evolution is further support for the new theory. This conclusion is difficult to accept because it would indicate that the gene pool could hide an almost limitless variety of potential form but the theory assumes that the hidden variety within the gene pool is a more narrow range around the form which is most commonly expressed in the phenotype of the organism. Parallel evolution therefore is either the result of a similar response to a similar need or perhaps the similarities between species is illusory in the sense that our ability to discern the differences between species is handicapped by the presence of one or two similar characteristics. A marsupial wolf may look like mammal wolf because it has a few important characteristics in common but if the comparison were made on an organ by organ basis using some quantitative technique then I think that the two wolves may prove to be as different as a dog and a cat. After all, the self-developing genome is trying to cope with a limited set of physical conditions which exist on the earth and when you consider that the total number of species which have ever existed is more than a billion then it is hardly surprising that evolution occasionally comes up with repeats. Again, the faster rate of evolution predicted by the theory of the self-developing genome makes such coincidences more probable. Darwinian random mutations have little or no chance of coming up with useful structures once never mind twice.

Metamorphosis

Although not directly relevant to the theory of the self-developing genome it is interesting to note that an individual genome itself is capable of storing different forms. One of the most extreme examples of this type of storage is the ability of certain organisms to undergo a complete metamorphosis during their life cycle. The two best known examples of metamorphosis are the transformation of a caterpillar into a butterfly and the equally amazing change from a fish-like tadpole to an air breathing frog. One of the most interesting examples is that of the crustacean Sacculina which is admirably described by

Michael Denton in his book *Evolution: A Theory in Crisis*; "The egg hatches into a typical free swimming crustacean larva, which then develops a bivalve shell and comes to resemble a small water flea. During this stage the larvae develop an organ for piercing the integument of a crab. On entering the crab it undergoes one of the most extraordinary transformations in nature. From being a crustacean-like organism it gradually changes, losing all its internal structure and organs, into an amorphous mass of cells which sends out root-like processes into the tissue of the crab. These processes, which resemble fungal fibres, ramify through the crab tissue absorbing nutrients and convey them back to the main mass of the organism which at this stage is little more than an egg producing bag."

These are not only examples of different phenotypes stored in the genome but also demonstrate the hierarchical structure of the genome with some genes taking overall control of the timing and coordination of the metamorphosis. Metamorphosis is not directly relevant to the new theory because it does not demonstrate the kind of stored potential which the new theory predicts but it is an example of the sophisticated programming ability of the genome. It also makes the transformation over geological time from fish into amphibian more understandable when you consider that a similar transformation happens within the lifespan of every frog.

Hidden Form

In his book *The Great Evolution Mystery*, Gordon Rattray Taylor makes much of the type of hidden form which can be stored within an individual genome. He gives the name "masking theory" to the idea that blueprints for building phenotypes can be hidden for millions of years before suddenly being expressed by the species. Gordon Rattray Taylor died shortly after completing his book and I often wonder if he would have maintained and developed his interest in alternatives to Darwin's Theory. He comes very close to postulating the theory of the self-developing genome and the following lengthy quote from chapter thirteen of his book will serve to make the point that I am not the first to suggest such an idea; "There are evidently things about the functioning of the genome which we do not yet understand. What is the range of possibilities available to it? Its repertoire must, in any given species be limited. The logical deductions from the observable facts of evolution seem to me to be: (1) the genome must run through many permutations, discarding all those which are not inherently viable, and trying out the viable ones for effectiveness in the real world; (2) it must have a facility for storing such potential forms and for realising or activating them when circumstances demand a change. But in doing so it will eventually exhaust all the possibilities open to it, at which point its evolution will cease. The organism concerned will continue to flourish as long as conditions remain unchanged, but will become extinct if they change in a way to which the repertoire of stored variations cannot effectively respond.

These are bold conclusions. Some biologists still deny the possibility of storage, although the discovery of redundant DNA has made the first of the above assertions plausible. The postulated mechanism for release of variations is not known. However, a system of this kind would explain the changes in tempo, the sudden jumps which Eldredge has documented, the stasis of sharks and lungfish, the rapid evolution of man, and the appearance of atavism and anomalous structures with much else. The genome must have its own logic." Mr. Taylor is obviously suggesting that the genome itself is the store-house of different forms but my own feeling is that the gene pool and the interplay of dominant and recessive genes is sufficient to explain the mechanism of storage. However, Mr. Taylor's masking theory remains an interesting possibility and I would like to consider it further just in case he's right!

When Mr. Taylor gives numerous examples to support his masking theory he gives no indication of the source of these body plans and he puts forward the masking theory as a possible explanation for parallel evolution. He cites the example of mole-like creatures which have developed independently in Asia and South America and also anteaters which have evolved on four separate occasions. I cannot agree with his idea that these body plans were stored in some primitive common ancestor of these creatures. This idea would lead to the conclusion that a primitive genome had hidden within it a certain limited number of body plans which can be revealed at some time in the future. But without an explanation of the source of these plans such a notion is not very convincing.

Mr. Taylor is convinced that a type of experimentation is at work and when talking about the primitive creatures known as Echinodermata he says, "Many of these early creatures experimented with their entire body form, moving their feet or their neck, shifting their eyes (some had eyes in their tails) and then twisting and contorting themselves. Far from being a struggle to survive, it looks more like one glorious romp."

The idea that the genome contains the plans for innumerable different phenotypes is really not very revolutionary when you think about it. We have already mentioned the metamorphosis of tadpole into frog and in chapter three we briefly touched on the subject of embryology. We said that the phenomenon of embryology demonstrated that the genome could store an almost infinite variety of forms taking the organism from a single cell of DNA to a mature body through innumerable fully working, perfectly coordinated stages. We mentioned that the phenomenon of embryonic growth gave us clues to the methodology of information storage within the genome. There is obviously not a full set of genes for every infinite stage of embryology but we have to imagine a four dimensional storage system where the fourth dimension represents the plasticity of the organism from one working model to another working model just one small stage removed. When we eventually gain a full understanding of the genome we will see that the simplistic idea of one gene coding for one protein is only a small part of a program of incredible complexity and ingenuity for controlling both embryonic and evolutionary development.

Although biologists understand very little about the rules governing embryonic development surely they must accept that the genome holds the key to the whole process. If we accept that part of the genome is responsible for guiding, controlling and coordinating embryonic growth then surely we can accept that evolutionary developments can also be planned and executed by genes; it is certainly not beyond the realms of chemical possibility. If we accept without question that a growing embryo has a built-in plan for development why can't we accept that the linked phenomenon of evolution also has a built-in aim.

The parallels between embryology and evolution have often been noted by biologists and the remarkable resemblances between different species when compared at early stages of embryonic development have led to the suggestion that an organism traces its evolutionary route during its growth to adulthood. The human embryo for example develops gills and tails en route to becoming more human-like. The idea that embryonic development is a reflection of evolution is too simplistic but there are many examples of primitive characteristics being retained by an organism long after that particular characteristic has ceased to be useful. Whales retain the vestiges of the legs which they once needed as land mammals, moles have the remnants of eyes even though they have no use for them and our own wisdom teeth are another example of a primitive characteristic. Sometimes a characteristic which we think has completely disappeared makes an occasional reappearance. This phenomenon is known as an atavism and the best known example is when a horse is born with three toes instead of the customary one. The ancestor of the modern horse, of which there are many fossilized specimens, can be seen to have had three toes.

Darwin himself used the idea of rudimentary organs as evidence of common descent. In *The Origin of Species* he says, "Organs or parts in this strange condition, bearing the stamp of inutility, are extremely common throughout nature. It would be impossible to name one of the higher animals in which some part is not in a rudimentary condition."

The idea of characteristics and features fading in and out of phenotypic expression fits very well with the new theory of the self-developing genome. The new theory expects the gene pool to be a store of variation around a mean with extreme variation being less available for expression than variation which is closer to the mean. A characteristic is not stored on an all or nothing basis but it is more a matter of degree. The possibility of producing more or less of a whale's leg can be visualized by a bell-shaped normal distribution curve. For example, the whale's leg which is most often expressed in the phenotype would be represented by the peak of the curve. Other types of leg would be expressed less often. The direction of change generated by the self-developing genome will be chosen on a random basis and this factor could cause the curve to drift either to the left or right. Natural selection, by fine tuning the organism to its environment could also cause a directional shift in the position of the curve.

Another way of looking at vestigial structures is to think of embryonic development as not continuing along its full path for a particular characteristic but stopping short before full development and leaving only a rudiment as evidence of a once useful structure. Evolution has occasionally made greater use of restricted embryonic development in a phenomenon known as neoteny. This

82

process has sometimes led to evolution dispensing with the adult stage of embryonic growth in a situation where the juvenile version of the organism attains sexual maturity. The most quoted example of neoteny is the case of the axolotl, an amphibian which is spawned in water and then develops through an aquatic tadpole stage to become a land based air breathing adult. There is however a species of axolotl in Mexican lakes which does not metamorphose into the adult stage but which remains and reproduces at the tadpole aquatic stage. The most interesting example of neoteny concerns man himself. There is general agreement amongst biologists that the adult form of Homo Sapiens is closely related to a juvenile ape. Such characteristics as lack of hair, brain size in relation to body size, shape of skull and relative proportions of trunk and limbs all point to the adult Homo Sapiens being more similar to a juvenile ape than to an adult ape. Evolution is just as capable of eliminating a previously expressed form as it is of expressing a new form which was previously hidden. The self-developing genome is experimenting not only with different versions of finished bodies but also with the various stages of building these bodies.

Trends

The tendency for animals to get bigger was mentioned in chapter two as an example of the weak link between environment and organism. In fact some biologists have noted many other trends in evolution which seem to have continued beyond the point of usefulness. One of the classic examples is that of the Irish Elk which developed gigantic antlers before becoming extinct. The disadvantages of such large antlers in terms of the accepted norms of natural selection greatly outweighed the advantages. An exaggerated feature of an organism such as peacock feathers or the tail of the lyre bird are explained as being developed by sexual selection. This occurs when the female (usually) is attracted to some particular feature of the male and the more the male possesses that feature the better its chances of attracting a mate.

Gordon Rattray Taylor in his book *The Great Evolution Mystery* makes much of this tendency to "overshoot" as he puts it. He describes the case of the extinct oyster Gryphaea which is initially found in the fossil record with a 10 degree coil but then continues to increase its coiling until it reaches 540 degrees millions of years later. Such a trend has never been adequately explained in Darwinian terms, neither has the teeth of the extinct smilodon which in Taylor's words "became so large that they could not close their mouths."

These are examples of the exaggeration of nature which are quantifiable and easy to measure but it is not so easy to put a value on complexity and if the tendency to continue in a certain direction could be applied to complex structures then the high degree of complexity achieved by nature could be explained. I am still undecided as to the significance, if any, of these apparent trends but I must admit that initially I felt that this phenomenon needed an explanation in terms of the self-developing genome. If it could be demonstrated that changes in the environment occurred in a trend-like fashion e.g. gradually getting colder, then I initially thought that it would be reasonable to expect the developing genome to follow such a trend. Keeping up with the changing environment would indeed be a great advantage and the unconscious knowledge of direction would be gained by the negative feedback of natural selection. (The term "negative feedback" will be used to denote the tendency of natural selection to eliminate less useful characteristics). A knowledge of evolutionary direction contained within the genome is not out of the question as we have already noticed how the previous evolutionary history (and therefore trend) is partially demonstrated during embryonic growth. The idea that the self-developing genome could follow a trend is only a small stage removed, in terms of chemical possibility, from the pure theory of the self-developing genome which predicts that evolutionary direction is random and trends would only be achieved by the vagaries of random drift or natural selection.

If the ability to evolve in a certain direction was a characteristic which could be inherited then this would have the effect of greatly speeding up evolution. If there was a selective advantage to big bodies then the selection of a trend to develop big bodies would be a much quicker route to big bodies than both the accepted route of natural selection and the new route of the self-developing genome. Evolution would favour any such "speeding-up" mechanism if there was an advantage to it. In chapter three it was stressed that the self-developing genome operated on a random basis and if this methodology creates sufficient variety to cope with an unpredictable environment then the idea of the inheritance of trends might well be unnecessarily complex.

Although the inheritance of directional trends might be unnecessary, the *degree* of experimentation which determines the rate of evolution is more likely to be an inherited characteristic. The inheritance of different rates of evolution would go some way to explain why some organisms do not change their appearance for millions of years whereas others appear to evolve more rapidly. The evidence of the molecular clock indicates a constant rate of evolution when applied to genes which code for particular proteins but the results from protein sequencing clearly indicate that each protein evolves at its own particular rate. The whole question of differential rates of evolution must remain unanswered and we can only guess at the exact rules which control the self-developing genome.

The usefulness of a built-in trend would depend on whether the environment changes in a trend-like manner but I think we can safely assume that if any directional changes *do* occur they are not significant enough to justify a genetic mechanism to follow them. We know, of course, that long term trends exist in such things as weather patterns; ice ages come and go; but weather is only one of dozens of constituents of an organism's environment. We know that it is extremely difficult if not impossible to predict the outcome of a situation when many independent forces are at work which can affect that outcome and therefore I think we must conclude that the genome is trying to cope with a randomly changing environment and therefore its best chance of coping with such an environment would be to also develop on a random basis.

This leaves the mystery of trends unsolved. They are possibly just another example of sexual selection in action; bigger bodies are seen as more attractive. On the other hand trends could be merely an insignificant consequence of the suspected link between embryonic growth and evolutionary development; after all, bodies tend to get bigger as the infant grows into adulthood.

Sexual Reproduction and Death.

There are two aspects of life and evolution which Darwin's Theory has always had difficulty explaining. These are the twin mysteries of sexual reproduction and death. We might expect that sexual reproduction would not be favoured by Darwinism because, for example, the genes of the female have only a 50% chance of being passed on by the offspring via sexual reproduction whereas all of her genes would be passed on to the future if she reproduced asexually. And similarly if survival of the genes is the aim of evolution then why do organisms die? Why don't they live forever? There should at least be some organisms which have come up with everlasting life as a useful survival strategy. The modern explanation of the popularity of sexual reproduction in evolution says that the mixing and rearrangement of genes which takes place during sexual reproduction is an extremely important way of keeping ahead of rapidly evolving viruses. A virus is likened to a key and cells of an organism are seen as having a lock. The virus can only enter and hi-jack the cell if its key fits. Sexual reproduction has the effect of continually changing the lock every generation. From what we know of the ingenuity of life on earth surely evolution would have come up with a simpler way of protecting organisms from an invading virus than the incredibly laborious and inefficient method of sexual reproduction.

Death also fares badly under Darwinism. The two explanations most recently put forward run something like this. Firstly, it is a very expensive thing, in terms of design, effort and raw materials, to produce a living body. It would be even more expensive to produce a body that could last forever. We do not build cars to last forever, not because we do not have the technology, but because they would be too expensive. What is the point in building a car to last for a thousand years when there is an excellent chance that it will be damaged beyond repair in an accident within say thirty years? Exactly the same argument is put forward by Darwinists. Why should evolution go to the trouble and expense of designing an organism to last for a million years when it is likely to die in an accident or be eaten by a predator within a few years. The second reason put forward for death is the accumulation over time of copying errors when cells which make up the organism reproduce themselves. An ageing organism does indeed show signs of deterioration over time. Gradually every part of the body starts to malfunction and the accumulation of copying errors seems a reasonable explanation. But if the survival of the genes is the fundamental concern of evolution then surely evolution would have come up with a more fool-proof mechanism for repairing or rejecting cells containing copying errors.

We have seen that the self-developing genome gives evolution an unconscious aim of designing and producing new working models. This mechanism would be quite impossible if the model was never scrapped. If the body lived for ever then change could only come about by the ongoing modification of the living body. This would be extremely difficult if not impossible. When man wants to improve the design of a machine he knows that eventually he will have to build a new machine from scratch rather than try to modify an old model. We did not evolve the modern jet plane by adding bits and pieces to the original bi-plane. The same rule applies to any engineering project; you always start again with new blue-prints. This is another example of modern technology mimicking what nature has been doing for eons. Don't try and amend an old model; scrap it, redesign it and start again. Death and rebirth are vital ingredients of creating variety. Evolution would be almost impossible without this cycle of life. What about sexual reproduction? Again it is a vital part of the process of producing variety. The unconscious aim of evolution is the production of variety and sexual reproduction allows an individual organism to mix its genes with any other individual member of the opposite sex. The potential for new variety is enhanced incredibly compared to the alternative of asexual reproduction in which you start off with just your own genes. With sexual reproduction the concept of the gene pool comes into play. All the genes of a reproducing species can be seen to be available to the self-developing genome as its raw material for evolution. Evolution can now be seen to be a three part mechanism; death, sexual reproduction and the self-developing genome. We will return to this idea in the final two chapters.

Chapter Five

The Origin of Species Revisited

The fundamental difference between Darwin's Theory and the theory of the self-developing genome is that the new theory predicts that evolution is now a process which proceeds automatically and in the short-term is largely uninfluenced by events in the outside world. The parameters which limit evolution to forms which will be useful in the conditions met on earth were initially programmed into the genome by the negative feedback of natural selection and are subsequently reinforced in the same way.

On closer examination of *The Origin of Species* the reader is struck by two observations. Firstly, there has never been, and might never be, such a thorough review of nature, so successfully summarised in one volume. It is a work of diligent scholarship, where a painstaking collection of facts and observations is combined with a cool analytical interpretation of the evidence to reach reasonable and logical conclusions. The reader can have great faith in the conclusions drawn by Darwin and it is only with the enormous benefit of twentieth century science that we are able to question some of those conclusions.

The second observation is that Darwin constantly takes a cautious and moderate view when acknowledging those aspects of the mechanics of evolutionary change of which he was ignorant. He gave entire chapters in his later editions to pointing out the weaknesses and objections to his theory. He fully accepted that he had little knowledge of the mechanism of inheritance or the rules regarding the source of variation both of which were crucial to his theory. Experiments had in fact been conducted during Darwin's lifetime demonstrating the fundamental laws of inheritance but sadly this important research never came to his attention. The idea of discrete genetic units of inheritance helped to fight off one of the most valid criticisms of Darwin's ideas. Darwin assumed that an offspring was in some way an average of its parents in terms of physical characteristics and this idea of "blending inheritance" was jumped on by Darwin's critics because they quite correctly said that such blending would dilute any new advantageous feature out of existence. The new feature would therefore never be available for natural selection to increase its frequency. For many biologists the discovery of genes finally convinced them that Darwin was right as the problem of blending inheritance was now resolved.

The other important feature of evolution about which Darwin admitted his ignorance, was the source of variation on which natural selection could act. I doubt whether Darwin would have been very impressed with the idea of random mutations being the only source of variation. In *The Origin of Species* he says, "I have hitherto sometimes spoken as if variations were due to chance. This, of course, is a wholly incorrect expression, but it serves to acknowledge plainly our ignorance of the cause of variation."

Darwin in fact thought that the continual use or disuse of a particular organ could be a source of variation in a manner which is similar to Lamarckian inheritance. Lamarck, a brilliant French biologist in the early nineteenth century, thought that a characteristic gained in your lifetime could be inherited by your offspring. For example, if you spent your life lifting heavy weights, then your offspring would tend to be born with bigger muscles. In the absence of any understanding of genes and the chemistry of inheritance this is an extremely sensible conclusion to arrive at. The idea that variation is solely a random happening did not cross the minds of such great thinkers as Lamarck and Darwin and yet modern day biologists accept it without question. The mystery of the source of variation is explained by the theory of the self-developing genome; the genome itself is the creator of the variation which natural selection can use to fine-tune the organism to the environment.

Selective Breeding

Darwin drew heavily on his knowledge of domestic animals and pointed out that selective breeding was merely an unnatural version of natural selection. He also pointed out, quite correctly, that rapid change was possible. This is another reason why he would have rejected random mutations as the source of variation; the speed with which a new characteristic in a domestic animal can be created is far in excess of the acknowledged speed of random mutations. I recall an article in the Daily Telegraph magazine which gave fascinating photographic evidence of the significant changes in the various breeds of dog brought about in just one hundred years of selective breeding.

It is commonly thought that there is a limit to the changes produced by selective breeding because it seems to become more and more difficult to develop a particular feature in a domestic plant or animal. This observation would fit very well with the theory of the self-developing genome because if intense selection took a particular feature to the extreme limit of the form stored in the gene pool time would then be needed for the further evolution of extremes around a newly established norm. The idea that new random mutations can account for a hundred years of successful dog breeding is difficult to believe.

Diversity.

The new theory recognises that the central theme of evolution is survival by means of creating variation. Darwin himself recognised the importance of diversity and the occurrence and persistence of variety is so evident in nature that it is surprising that biologists have not sought a

91

mechanism more dynamic than Darwin's Theory as an explanation for this variety. We not only see variety in the taxonomic arrangements of species, classes and phyla but it also manifests itself in the building blocks of our own bodies in a phenomenon known as polymorphism. The protein Haemoglobin for example occurs in about one hundred different varieties and the exact cause and purpose of such polymorphism has been subject to some vigorous academic debate. Polymorphism is merely another example of nature seeking variety and hanging onto variety when it is not disadvantageous to do so.

Coordination

One of the most important predictions of the theory of the self-developing genome is that new versions of fully coordinated working bodies are created by the genome before being exposed to the outside world. This notion overcomes the untenable claims of pure Darwinists that random mutations can develop a body which contains millions or billions of interconnected, perfectly proportional parts. Darwin himself was worried by this miracle of coordination and he assumed that there must be some underlying mechanism which accounted for this phenomenon. He suggested a "law of correlated variation" to account for this coordination and by this he meant "that the whole organization is so tied together during its growth and development, that when slight variations occur in one part and are accumulated through natural selection, other parts become modified. This is a very important subject, most imperfectly understood, and no doubt wholly different classes of facts may here be confounded together."

As always Darwin recognised those parts of evolutionary theory which needed further explanation. In another chapter he reminds the reader again of his lack of knowledge "it should always be borne in mind that when one part is modified, so will other parts through certain *dimly seen* causes such as correlation." (emphasis added).

Our knowledge of genetics has had little impact in understanding the laws behind building bodies although we do know that genes become linked together in order to achieve coordinated action.

Missing Links

Although Darwin was adamant in his view that evolution is gradual he recognised that the evidence from the fossil record indicated that evolution might be a series of jumps, rather than a gradual process: "Geology assuredly does not reveal any such finely graduated organic chain; and this, perhaps, is the most serious objection which can be urged against the theory." The new theory of the self-developing genome provides a solution to the problem of the jumps in evolution. At the molecular level evolution is smooth and continuous, but phenotypic evolution is not smooth because the pressure from natural selection can either hold back or speed up the expression of the forms which are constantly being created by the genome.

Speciation

Despite the fact that Darwin's great book was entitled *The Origin of Species*, the actual origin of a new species remains a mystery to this day and the whole subject matter of speciation forms the basis of a healthy debate in biological circles.

Darwin did not feel strongly about the need to define speciation because he thought that nature had no hard and fast dividing lines and he assumed that an individual member of one species could be linked to an individual of its closest related species through a graduated continuum of individuals, where each individual in the continuum varies imperceptibly from its neighbours on either side of it. The closest related species might well be extinct and so we do get considerable missing links in the continuum but in Darwin's view "Natura Non Facit Saltum," or "nature does not make jumps," and therefore the missing links will have existed at some time in the past. To Darwin, the creation of a new species from the original parent species was merely an extreme example of the diversity which is possible within the parent species.

Taxonomists sometimes give support to Darwin's idea when they split down one species into several sub-species. Although these sub-species may differ in some way from each other the different sub-species may still be capable of inter-breeding in certain circumstances. Modern theory, however, tends to favour more distinction and the missing links in evolution are now recognised by new theories which predict very rapid creation of new species followed by and preceded by relatively slow evolution.

Darwin's insistence on the smoothness of evolution has been vindicated when viewed at the molecular level where we find for example that individual human beings can be remarkably similar to their nearest primate cousins. Despite the pronounced differences of physical appearance between ourselves and chimps, it is occasionally possible to find two humans whose genes vary by a greater degree from each other than the amount by which one of them varies from a chimp.

Studies at the molecular level reveal the continuum which Darwin expected and the degree of difference between geographically isolated populations within the same species supports Darwin's view that one species fades into another in a graduated series of imperceptible steps. What we see at the morphological or phenotypic level is deceiving in that the differences at the molecular level are not exactly reflected in the phenotype and often a minute genetic difference can lead to a more significant phenotypic difference. For example, it is not evident by comparing the physical differences between chimps and humans that we have more than ninety-nine and a half percent of our DNA in common.

The theory of the self-developing genome tells us that evolution proceeds within the genome without continual reference to the outside world. The limitation of this "blind evolution" is controlled by the negative feedback from natural selection. The analysis of protein sequences indicates that evolution is proceeding at a constant rate and if we assume that the genes responsible for giving the organism its large scale physical structure (as opposed to its protein structure) are also evolving at a constant rate then we must assume that our old friend natural selection is disguising the smooth evolution which occurs within the genome. The assumption that evolution of protein structure and evolution of body structure and form are proceeding at the same rate is supported by the fact that hierarchical patterns of species classification as revealed by protein sequencing correspond very well with the traditional taxonomic relationships worked out by biologists based mainly on body structure. The evolution of proteins as revealed by protein sequencing appears to be truly random. The same physical protein can consist of different amino acid sequences without affecting its characteristics and the variations in the sequences of a particular protein do not appear to have any adaptive significance. Of course a random change in the amino acid sequence which destroyed the protein would be eliminated by the negative feedback of natural selection. I see no reason to think that evolution of form or structure proceeds in a different manner; what we see in nature is a random expression of form and the comfortable fit between a species and its environment is due both to the feedback of natural selection fine tuning the organism and to the continual search undertaken by the organism for a more comfortable niche.

The pronounced differences of form between two closely related species which are often greater than would be expected from the evidence of protein sequencing could be due to the methodology of building bodies or embryonic growth. Any minor change to structure at an early stage of embryonic growth could have an exaggerated knock-on effect at the later stages, especially in the situation where a later stage is relying on an earlier stage having been completed. Thus a minor genetic change at an early stage of the embryonic growth plan could have a dramatic effect on the phenotype. We have already noted the possible connections between evolution and embryonic development and I strongly suspect that these two aspects of life will prove to be more inter-linked than we currently suspect.

The question of species formation was left unanswered by Darwin and has remained controversial to this day. Biologists are typical scientists in that when looking for a rule to explain a phenomenon such as species formation they would prefer to have one law to which they could give a name and which would account for all the observed events. The search for the golden rule to account for species formation has led to many suggestions and it is my experience that when intelligent people all claim to have the correct answer then they are either all wrong in their conclusions or there is an answer at a higher level which allows all the lower level answers to be correct in a limited set of circumstances. The five major arguments can be summarised as follows:

1. Does speciation occur only when a population gets geographically separated from the main group or can speciation occur without physical isolation?

2. Is speciation the result of a major genetic mutation which creates a barrier to sexual reproduction or is such a barrier created by a physical change?

3. Is the gap between one species and its nearest related species always imperceptible or does nature take big jumps?

4. Is a new species created because of environmental pressures or is there no adaptive significance to the new species?

5. Is a new species formed by the splitting of a parent species into two groups or are only a few isolated individuals capable of creating a new species?

At least everyone is agreed that a species can be defined as a group which is reproductively isolated from all other groups but it is the method by which the group achieves this isolation which is the subject of disagreement.

Well, what does the theory of the self-developing genome predict? Simply that there is no single rule because the real creator of new species is the self-developing genome itself and speciation will be achieved by any available route in any conceivable manner. Despite the restrictions placed on the self-developing genome by natural selection the pressure to diversify will eventually overcome these restrictions and the mechanism of sexual reproduction will inevitably lead to the creation of a new species.

Of all the routes taken by evolution to speciation my own personal favourite is one which is seldom considered by biologists, namely a change in behaviour by part of a population which could lead to reproductive isolation. We know for example that some closely related species differ only in courtship behaviour or in the timing of their mating season and this differentiation is directly related to reproductive isolation. But *any* change in behaviour could lead to speciation because of the strongly instinctive tendency to imitate as seen in offspring. Let us say for example that a bird develops, via the self-developing genome, a skill to hunt in a particular way. The offspring of that bird, and indeed other unrelated birds, might try and imitate the parent and if they are not well equipped to develop the skill they would be at a selective disadvantage compared to those who can develop the skill. If, like the peacock's feathers, birds with the new skill were sexually preferred then the skill could quickly spread and reproductive isolation could follow. Similarly, the skill of bipedal locomotion might have started by the action of one individual followed by imitation by others leading to reproductive isolation and bodily changes via the self-developing genome and natural selection to improve the skill. The reason why whales returned to the sea could be nicely explained by a change in behaviour. If, for example, a land based ancestor of the modern whale developed a more oxygen efficient haemoglobin via the self-developing genome, then the propensity to wander into new environments, which is common among all animals, might have led the whale's ancestor to try deep sea diving instead of just splashing about in water. This practice could lead to imitation and therefore the elimination of those who lacked the new haemoglobin.

Darwin recognised the part played by behaviour in his theory when he said "As we sometimes see individuals following habits different from those proper to their species, we might expect that such individuals would occasionally give rise to a new species, having anomalous habits and with their structure modified from that of their type. And such instances occur in nature."

Geographical isolation could not possibly be responsible for creating the billions of species which we think have existed but we know for a fact that behavioural differences allow many different species to share the same geographic area and therefore the evidence would suggest that behavioural differences should be given more importance by biologists when discussing speciation.

The theory of the self-developing genome would predict that a new species is created before natural selection plays its part in fine-tuning the new species to its environment. Darwin's Theory however, would predict that adaptation occurs first because it is assumed that species creation (or evolution) is caused by natural selection. The fact that most biologists agree that speciation occurs before adaptation doesn't really make sense when you consider that they mainly Darwinists.

The vast number of species which are known to have

existed is evidence itself of the natural desire which evolution has to diversify but the question which the reader may well be asking is why do we have species at all? Why are organisms not part of one large species? Let us imagine for a moment the existence of just one species inhabiting every corner of the world. Organisms might have some local adaptations to cope with local environmental conditions but every organism would be capable of mating with every other organism. The rules of mathematics being what they are I think that it is a fair assumption to say that all members of this single species would, by consequence of their common gene pool, be remarkably similar. There would be some variety but the creation of variety would be very restricted. We saw in chapter four that the mechanisms of sexual reproduction and death are needed by the self-developing genome in order to create variety and we can now see that species formation also plays a vital role. Species formation is crucial to the unconscious aim of evolution, the creation of variety. The formation of separate, isolated gene pools gives evolution a new starting point to create more variety. The direction of change in this new gene pool is totally uninfluenced by the direction of change of the parental species. We now have the final piece of the jig-saw, the final part of the overall mechanism of evolution which ensures that the random search undertaken by the self-developing genome will eventually take us from simple organisms to man

Chapter Six

Further Speculation

Educated and knowledgeable people held onto the notion that the earth was the centre of the universe for more than a hundred years after Copernicus proposed his view of the earth in daily motion about its axis and in yearly motion around the sun. The traditional view was held onto despite astronomical evidence to the contrary. The traditional view was precious to humanity who regarded themselves as chosen by God to live in a universe designed around them. Today a similar situation exists with regard to the establishment view of evolution. The scientific community is desperate to hold onto neo-Darwinism because the alternatives are not as tidy. The alternatives require more of a purpose or force in evolution. Such a force is anathema to Darwinists. Natural selection is now such a popular notion that it is applied to not only genes and organisms but to other evolving systems such as business organisations and economies. The phenomenon of natural selection requires neither purpose nor direction but change results from the struggle to exist which occurs between competing units. Natural selection has turned out to be one of the most useful scientific concepts ever formulated but it cannot, on the grounds of mathematical improbability, be responsible for designing complex biological organisms.

The idea that innumerable random changes in the genome can accumulate over time to eventually construct an organism as complex as the human being is inconceivable on the grounds of improbability. Biologists have convinced themselves that natural selection acting on this highly inefficient random mechanism changes the situation from being a highly improbable event to an event within the realms of possibility. I find it very difficult to get onto their wavelength and to understand their reasoning. Random changes in the genome do not easily produce anything but nonsense at the level of the phenotype and an innumerable series of well coordinated changes would be needed to produce the slightest stable change. The route from the initial program contained within the genome to the synthesis of a fully working organism is so complex that we are still only scratching the surface in our attempts to understand this phenomenon. Single favourable mutations would effect only one stage in the chemical manufacturing process and several if not thousands of coordinated changes might be needed to eventually achieve a favourable change in the finished product. In chapter two we have argued against the possibility of even a single favourable mutation. Therefore what chance is there of the random process producing the numerous changes needed to make an impact on the phenotype? And what chance does a random system have of producing the numerous batteries of inter-dependent and coordinated changes necessary to produce significant structural changes to organs as complex as the eye and the brain?

Perhaps the rare single stage mutational changes as would have been needed in the early days of primitive DNA can be envisaged to have happened as a random event but to achieve the vast array of coordinated changes which have subsequently occurred would require a mechanism which is less reliant on random events and therefore more probable. In the previous chapters I have proposed such a mechanism and have described it as the self-developing genome. The compelling argument which initially led to the idea was that given the evidence of the seemingly boundless complexity of chemical possibility which we observe in nature and given the necessity for the organism to be flexible in its approach to the environment then life would have come up with a mechanism which was an improvement on the incredibly inefficient reliance on random mutations.

The theory of the self-developing genome turns the accepted evolutionary theory on its head. The traditional Darwinian view is that the genome is trying its best to pass on faithful copies of its genes to the future but the self-developing genome, on the other hand, is trying to produce variety. Under the pressure of the self-developing genome natural selection is more an instrument of restricting change rather than a promoter of change. The degree of evolution achieved at the phenotypic level in any given period of time is held back by natural selection, but this relentless production of variation will eventually lead evolution down every conceivable survival route.

Evolution can now be seen to be a random search moving one way or another, a pendulum swinging to and fro through time. The unconscious aim of evolution is the creation of variety and the random search for different forms coupled with the mechanisms of death, sexual reproduction and speciation will give evolution the ability to progress from simple organisms to human beings. This progress is an inevitable consequence of the mathematical rules of evolution.

We no longer have to conjure up convoluted reasons to explain the adaptation of every organism to its environment. With the self-developing genome the order of events is reversed; evolution occurs first and then the organism finds a niche in which it is most comfortable. Natural selection still plays a part in fine tuning an organism to its environment but if natural selection can be said to be significant in evolution we have to go back to the manufacturing or design stage and here we find that natural selection has played its part in creating an efficient mechanism for the creation of variety.

The new theory explains why life appears so randomly spread. Life appears to be capable of taking on any shape and size within the limits of what is chemically possible on our planet. The behavioural systems of organisms also seem to have evolved under the influence of the "whatever is possible" rule. If we accept the new theory we do not have to explain each and every life form in terms of its adaptation to the environment.

The new theory explains why the availability of niches plays a part in evolution. The genome is bursting to evolve and new species will be created when a few members of the old species are not restricted by a tight fitting environment.

The mystery of coordinated changes is also explained by the new theory. The algorithm behind the self-developing genome is a program which attempts to create bodies which actually work. The program has been designed by natural selection to create good working models that can survive.

The creation of new species has always been a source of controversy between Darwinian evolutionists. The new theory says that all routes to a new species are possible but the driving force behind the creation of a new species is the automatic production of variety within the gene pool. Occasionally, this build up of variety bursts through the prison of the sexually isolated species to create a new separate species. For the same reason the new theory explains the mystery of rapid evolution and the apparent gaps in the fossil record.

The new theory predicts that variation accumulates in the gene pool without being immediately vetted by the environment and this prediction fits well with the evidence from selective breeding. The only explanation of the fact that we can very rapidly alter the characteristics of an organism is that the mutations called upon were already in the gene pool. We know that changes produced by selective breeding eventually slow down or run out and the new theory would predict that time is then needed for new variation around the mean to be generated within the gene pool.

Chapter four argues that the new theory fits very well with the evidence. I cannot pretend to understand the exact mechanism of the self-developing genome but it is the principle that a less than completely random system is contained within the genome which is the important point. The method by which the genome reduces the odds in favour of evolution in the lottery of life is likely to consist of a complex multi-hierarchical process which will astound us with its ingenuity and the basic idea of the self-developing genome might well turn out to be too simplistic. I do believe that the new theory is far nearer the truth than Darwin's Theory and furthermore, I believe with absolute confidence that the fundamental concepts of Darwin's Theory are erroneous.

I can almost hear the groans of disbelief emanating from the scientific establishment. Pure Darwinism is a simpler explanation. The new theory is almost giving a purpose to evolution. The new theory boldly states that progress is inevitable. If the original self-developing genome was pre-programmed in such a way as to make progress inevitable then we must again return to the most important unanswered question in evolutionary theory. How did it all begin? I said in chapter one that I would return to this puzzle. Can we ever know the answer to this question?

We are still at the early days in our endeavours to unravel the mysteries of life and the universe and it is probable that our knowledge of the universe and everything which it contains is almost insignificant when compared to a theoretical total knowledge to which we aspire. How arrogant to suppose that our present generation is the one which is the first to understand the mystery of everything which we observe. History clearly shows us that evolution is a slow process and even a million years is a relatively short time in evolutionary terms. It is most unlikely that the present generation has just reached that particular threshold of intelligence to enable it to understand our own origins and the workings of the universe. If one single generation is all-wise then what is left for the billions of generations to come.

Not only the history of science but also our own personal life histories are littered with examples of jumping to the wrong conclusions because of limited knowledge. In fact all of our opinions are based on an imperfect set of facts and science suffers from exactly the same weakness. Lord Kelvin, the eminent nineteenth century physicist rejected Darwin's Theory to the end of his life because he was 100% confident that the earth was less than one hundred million years old. Darwin's Theory required a much older earth and Kelvin's confidence was based on his accurate calculation of the rate of heat loss from the initially white hot cloud of dust which formed the earth. His mistake lay in his ignorance of radio-active materials at the centre of the earth which continually create more heat.

I am often frustrated at the thought of our limitations. What knowledge and understanding is beyond our reach? If we recognise the limitations of the perceptual and intellectual processes of the lower animals then we must recognise that another 10 million years of evolution will prove that twentieth century Homo Sapiens was also extremely limited. We can only speculate on what the future holds for evolution and in many ways we are only speculating on what happened in the past. A glimpse in either of these directions is an experience which unfortunately is not available to us. We are firmly locked in a minute portion of space-time and we are constantly searching for clues which will help us understand regions beyond our reach. We are scratching the surface of our three dimensional space prison and although we are only now beginning to venture out into the solar system we hold out the hope that someday the vast expanses of time and space will be within our reach.

The previous paragraph recognises our limitations both in terms of what we do not know and in terms of what we are not capable of achieving but this does not prevent us from speculating on the past, the present and the future. My own speculations are concerned with evolution on a scale which is not limited to the narrow range between a simple organism and the human being. This book has been concerned with this narrow range because I wanted to discuss an academic subject on the home ground of academics. Darwinian biologists do not normally venture from the territory of living organisms and it is perfectly valid that scientific investigation is compartmentalized in this way in order to gain a more detailed understanding of a particular field of knowledge. In fact scientific research is becoming more and more compartmentalized because of the vast amount of detail involved in every category. Perhaps someday there will be specialist academic subjects concerned with linking together disparate areas of knowledge in order to seek patterns which only emerge with an overview.

A scientific overview was once the domain of philosophers. They were the thinkers who could absorb all scientific facts and attempt to construct generalizations and patterns from their elevated vantage point. However, this is no longer possible. An understanding of a single branch of physics can now occupy a scientist for half a lifetime. What chance does a single individual have of absorbing a dozen different fields of knowledge and then distilling all these facts in order to draw general conclusions?

Despite all these reservations it may well be sometimes useful to speculate on the broad picture even though these speculations may be little more than guess-work. My own personal speculation concerns the connection between life on earth and the origins and the future of the universe. I believe that the twin processes of the universe in general and life on earth are somehow connected. The existence of both the universe and life as we know it are unlikely events and I have often puzzled at the number of prerequisites which are necessary for the existence of life on earth. Of all possible temperatures which occur in the universe only a very narrow range will allow the basic chemical elements to react in a life-like way. It is the unlikely properties of these chemical elements which give rise to the phenomenon of life. One of these elements, carbon, has an atomic structure which just happens to give it the ability to combine with other elements to form chains of almost unlimited length.

106

There are after all only about ninety naturally occurring elements and we should count ourselves lucky that just one of them has the correct configuration to form the molecules of life. And what is carbon itself made of? Well, at school, we are told that it has a nucleus of protons and neutrons which is surrounded by electrons but at university we are introduced to the world of sub-atomic physics and we are told that protons and neutrons are in turn made of quarks. There are many other sub-atomic particles and forces which have been identified. The amazing fact is that each particle just happens to have the correct size and properties to enable atoms to exist; it is the properties of these particles which keep the wheels of the universe churning. To take a simple example of the importance of the existence of each building block, we can consider the ratio between the weight of a proton and an electron which from my schoolboy memory is 1840:1. If this ratio was just slightly different then the whole universe and the system of atoms contained therein would collapse. So what would happen if electrons were not that particular size? Would the primordial energy of the universe condense into something different, creating a world of different types of atoms and would there then be one particular atom like carbon that would facilitate the formation of life molecules?

Most biologists do not recognise the complexity of atoms. They often state that biology is concerned with complicated things such as DNA whereas physics is concerned with simple things such as atoms and galaxies. I only wish that I shared their knowledge of atoms so that they would also appear simple to me. The truth is that the synthesis of the simplest atom, hydrogen, which occurs in the centre of stars is an event as magical and as mystifying as the synthesis of the first DNA.

The most important event in our observable universe was its beginning which is commonly thought to have been a big bang. The initial rules or configuration at point zero set the scene for everything which followed. There are those who find it very easy to argue that given the initial rules then everything which followed including life on earth was an inevitable consequence. The initial rules must have allowed for the existence of electrons and quarks. Quarks would then inevitably condense into protons and neutrons which would then unavoidably link up with electrons to form atoms. Assuming that the particles of matter responsible for gravity were also part of the initial configuration then the large scale and chaotic universe would inevitably settle down to form galaxies, stars and solar systems. Of all the possible solar systems there would be some which contained planets which were at just the right temperature to allow the naturally occurring carbon atoms to form chains with other atoms which given enough random experimentation was bound to hit upon a structure which could duplicate itself which in turn could then lead to increasing complexity etc.,etc.

The question of the initial configuration of the universe is explored in Stephen Hawking's best selling book *A Brief History of Time*. Hawking points out that the initial configuration is only one of many possible configurations and we are currently observing the one configuration which led to us. Other sets of initial rules do not lead to us and therefore we are not there to observe them. This leads to the mind boggling possibility of an infinite number of other universes which we cannot observe.

It is this crucial point which we must take on board when discussing life's origins. We do not see the limits of creation in the observable universe. Cosmology is an almost purely mathematical scientific subject. We live in a three dimensional world of length, breadth and depth but cosmologists are convinced that other dimensions exist; dimensions which they can only see in the mathematics of the universe. Not only do other dimensions exist but travel between different parts of time and space and into other unseen universes is a theoretical possibility via black holes and worm holes. A superficial study of cosmology tells us that we are dealing with something far greater than our puny observable universe. 15,000 million years is an insignificant period of time compared to an infinite supply of similar periods of time in similar universes. I cannot accept that the DNA program for life happened by chance in the 15,000 million years of our universe but I am willing to be convinced that given infinite time and space then anything is possible.

The point of this speculation is that if the initial DNA program for life did not construct itself from inert chemicals on earth or even within our observable universe then given the possibility of an infinite supply of other universes we could still accept that it travelled through time and space and arrived on the earth's surface more than 3.5 billion years ago.

One of the three prerequisites which we had to accept in chapter one was that life started by chance as a chemical phenomenon which allowed inert chemicals to create a large self-replicating molecule. I touched on the complexity of the simplest known molecule of life and warned the reader that the occasion when life was first created was an extremely unlikely event. I deliberately understated the problem of how it all started in order to get on with the problem of how it all works, but it is this unresolved problem which has led many leading scientists to believe in panspermia or the seeding of life on earth from outer-space. Francis Crick, one of the discoverers of DNA, is one of these scientists who have drawn a blank on the enigma of life's origins and was quoted as saying "An honest man, armed with all the knowledge available to us now, could only state that in some sense, the origin of life appears at the moment to be almost a miracle, so many are the conditions which would have had to have been satisfied to get it going."

The whole concept of panspermia is really not inconceivable. We know from our own observations that life has this amazing ability to spring up everywhere and anywhere. Newly formed volcanic islands are soon teeming with life. Seeds are washed up as flotsam on the shore, bird's droppings contain indigestible seeds. We even find air bound spores floating miles above the earth's surface being transported to every corner of the globe. The genetic program can take on an extremely hardy and resilient form and under the right circumstances can remain dormant almost indefinitely until it finds suitable conditions to germinate and flourish.

Life could have arrived on earth on a comet. A rogue comet could conceivably travel from the far reaches of time and space linking up distant parts of the universe. The earth would have only needed one initial DNA program for life to have begun. The most contentious claim of the new theory of the self-developing genome is that the initial program *did* have an inevitable purpose. It was programmed to propagate intelligent life throughout the universe. It searched for a part of the universe with suitable conditions and it found them on planet earth. It achieves its aim of creating intelligent life by letting the mechanisms of death, sexual reproduction and speciation act on its relentless creation of variety.

If our evolutionary past is difficult to comprehend then what about our future? Can we foresee anything beyond the human species? The mammals of 50 million years ago would have had as little chance of comprehending Homo Sapiens as we will have of understanding life on earth in 50 million years time. Pessimists tell us that life on earth will have died out long before another 50 million years passes because of warfare or pollution. I cannot agree with this prognosis. My view of history is one of an accelerating improvement in our level of civilization interspersed with occasional steps backward due to growing pains and the upheavals caused by progress. So where will it all lead to? With the fantastic progress which we now observe in the space of a single century where will we be in a million centuries? Although the opportunity of travelling through time to the future would be too tempting to miss I do not think that we would feel very happy with what we would see. The next 100 million years of evolution will lead us down pathways which are currently unpredictable. The pace and direction of evolution will be inextricably linked to the product of that evolution although it cannot be said that the product of evolution will really be in control of his own destiny.

Clues to the future or at least the near future can be found today. Man is currently not the pinnacle of evolutionary achievement; there is now something greater than man, which although intangible in one sense is as real as any three dimensional organism. For want of a better expression, the pinnacle of evolution is currently the global culture or the pool of knowledge to which we all contribute in our lifetime but which outlives us and makes itself available intact to the next generation.

When observed from a different perspective a new pattern of evolution is discernible. For the first five chapters the story which we have considered occupies the narrow ground between simple organisms and humans. This is the home territory of Darwinism. It is on this territory that I wanted to do battle with Darwinists. However, when viewed from a cosmological distance then another broader pattern emerges. The energy-matter in the universe appears to be ratcheting itself upwards in a progressive climb up a hill that is leading to greater self-awareness, greater environmental control, greater knowledge and intellectual prowess. The pure energy created by the big bang which condensed into primitive sub-atomic particles soon led to the creation of the elemental atoms as we now know them. The creation of these atoms was a significant evolutionary step. It could be argued that an atom has quality and character which is worth far more than the sum of its individual sub-atomic parts. This phenomenon is repeated when atoms combined to form molecules. These molecules can be said give matter more varied properties and character. The third stage of the ratchet effect occurs when complex molecules get together to form DNA, the molecules of life. The chemical of life has character and ability far above the total sum of the characters and abilities of the individual chemicals from which it is constructed. The next ratchet upwards occurs when the molecules of life combine to produce an organism which has intelligence and self-awareness far greater than the sum of all the intelligence and self-awareness found in the cells from which it is made. The individual building blocks from which a human body is made are not aware of the higher level of intelligence which they have created. What is the next stage of evolution? Where will the ratchet now take us? The next stage of evolution is now upon us. It is the coming together of individual organisms, primarily man, to create pools of knowledge and intelligence. What started thousands of years ago with the transmission of culture from one generation to another is now manifesting itself into a global village. The communal pooling of knowledge is now encompassing the whole of mankind. We are more and more thinking as one enormous organism, the power of which is far beyond the sum of the individuals who make up the whole. The vast improvement in communications which we have witnessed is no meaningless technological sideline in evolution. It is a vital necessity to facilitate our inexorable climb up the evolutionary ladder. We are all part of the global coming together. We all contribute to the global intelligence. Our bodies die out but in our lifetime we contribute to the growing and evolving pool of knowledge. The global intelligence outlives us all and carries on through time as a new super organism. The global intelligence is a real phenomenon; it exists and is alive. We are as significant to the global intelligence as individual cells are to the human mind. The global intelligence will eventually leave behind the individuals who created it. It will eventually evolve beyond the comprehension of the individual organisms who transport

112

it through time. The complexity of the whole will become beyond the understanding of any one individual and we will become the equivalent of its cells. The global intelligence will continue to evolve and endeavour to explore its own origins. It will search the cosmos for the answers to those questions which our own intelligence is currently seeking. It might eventually link up with other global intelligences in other parts of the cosmos and we will then have yet another upward movement of the ratchet.

We can already see the beginnings of the evolution of the global intelligence. People are born and die with little or no understanding of the tools which they have relied on for the whole of their lives. We are nearly at the stage where our most sophisticated technology is not entirely understood by any one individual. It is only by coordinating the improvements to individual sub-units that improvements can still be made to the whole machine. I can easily envisage a new way forward for scientific advancement. Many scientists could simultaneously work on the same problem. They would have to sacrifice their chance of individual glory to the glory of the achievement of the group as a whole. Results from research or new ideas and thoughts would be typed into a database that could be instantly viewed by the other scientists in the group in other parts of the world who would immediately respond with objections and suggestions which in turn could be scrutinised by the others. It would be like several computers working in parallel; scientific advancement would be greatly enhanced.

Our past and our future will become clearer with the advancement of scientific knowledge. We are still guessing at this stage but speculation remains an important stimulus to research. The driving force behind an enormous number of research projects in the world of theoretical science is the desire to understand more about the universe at the one extreme and the chemicals of life at the other.

There are tens of thousands of scientists working in thousands of labs throughout the world looking into the far recesses of outer space or peering down electron microscopes delving into the molecular world of DNA and other organic chemicals. Of the two extremes I think the most significant results will come from genetic research. Research into the cosmos is now so purely mathematical that new discoveries only seem to effect those who can understand the numbers. On the other hand, genetic engineering can effect us all, especially in the field of medicine. Genetic engineering often gets a bad press especially when it conjures up visions of cloning and eugenics. What could genetic engineering achieve? Can we hope, for example, to eliminate all known diseases and therefore give everybody eternal life? The gift of eternal life would certainly have an amazing impact on evolution. One of life's great paradoxes is the fact that the human brain is at its physical peak years before it has had time to gain the experience needed to make full use of it properly. The actor John Gielgud bitterly complained that by the time he was experienced enough to play Romeo he was too old for the part. Unfortunately, the brain is already in decline when we have just managed to finish our period of training and as we gain more and more knowledge the brain becomes less and less efficient until eventually when our accumulated knowledge is at its peak we drop down dead and it's all wasted. Wouldn't it be wonderful if we could stop the ageing process at say age seventeen and then continue to pump knowledge into the efficient seventeen year old brain for the next hundred years? What power such brains could have. Imagine the contribution that these brains could make to the pool of knowledge. If the pool of knowledge does prove to be of importance to evolution then perhaps big brains will be the only significant feature which future humans will want to retain. You can conjure up some horrendous visions of the future by going down that route! Fortunately, I do not think we can easily control the ageing process by genetic

114

engineering. The ageing process is a fundamental part of the overall mechanism of evolution.

All this exhausting speculation does not help those of us who are seeking a meaning or a purpose to it all. Can we ever know the purpose? Perhaps that will be the knowledge which is always out of reach but it is the striving to find an answer which drives us to continue the search.

Do we actually need a reason for the existence of the self-developing genome? The answer is no, and the reason for the negative response is that it is only in retrospect that it seems that the original self-developing genome had a purpose. We would not be here to speculate on its past if it did not have the ability to produce us. The self-developing genome from which we evolved was inevitably going to lead to us. Every other possibility would have failed to create humanity. The universe and everything it contains ran through an infinite variety of possibilities until it eventually produced our own configuration of stars, atoms and the self-developing genome. All other configurations did not lead to us and therefore we can only observe the one history that did lead to us. Of course, this only takes the question of our existence one step further back. Why has the cosmos run through all these possibilities?

The answer to this question belongs to the future. We must accept the unknown for what it is and not be tempted to seek explanations in terms of unsatisfactory theories or notions. God is our ultimate refuge from the confusion which the unknown can create in our minds and He might indeed lie at the end of our search. But in the meantime let us leave Him as a possibility and continue to add to the pool of knowledge for the benefit of future generations.